Expert Systems in Finance

Throughout the industry, financial institutions seek to eliminate cumbersome authentication methods, such as PINs, passwords, and security questions, as these antiquated tactics prove increasingly weak. Thus, many organizations now aim to implement emerging technologies in an effort to validate identities with greater certainty. The near instantaneous nature of online banking, purchases, transactions, and payments puts tremendous pressure on banks to secure their operations and procedures.

In order to reduce the risk of human error in financial domains, expert systems are seen to offer a great advantage in big data environments. Besides their efficiency in quantitative analysis such as profitability, banking management, and strategic financial planning, expert systems have successfully treated qualitative issues including financial analysis, investment advisories, and knowledge-based decision support systems. Due to the increase in financial applications' size, complexity, and number of components, it is no longer practical to anticipate and model all possible interactions and data processing in these applications using the traditional data processing model. The emergence of new research areas is clear evidence of the rise of new demands and requirements of modern real-life applications to be more intelligent.

This book provides an exhaustive review of the roles of expert systems within the financial sector, with particular reference to big data environments. In addition, it offers a collection of high-quality research that addresses broad challenges in both theoretical and application aspects of intelligent and expert systems in finance. The book serves to aid the continued efforts of the application of intelligent systems that respond to the problem of big data processing in a smart banking and financial environment.

Noura Metawa is currently an Assistant Professor of Finance, Regis University, USA, and a Lecturer at the Faculty of Commerce, Mansoura University, Egypt.

Mohamed Elhoseny is currently an Assistant Professor at the Faculty of Computers and Information, Mansoura University, Egypt.

Aboul Ella Hassanien is a Professor at Cairo University, Egypt, Faculty of Computers and Information, IT Department, and the Chair of the Technology Center of Blind and Visual Impaired People.

M. Kabir Hassan is a Professor of Finance in the Department of Economics and Finance in the University of New Orleans, LA, USA.

Banking, Money and International Finance

For more information about the series, please visit www.routledge.com/series/
BMIF

Expert Systems in Finance

Smart Financial Applications in Big Data Environments

Edited by Noura Metawa, Mohamed Elhoseny, Aboul Ella Hassanien, and M. Kabir Hassan

LONDON AND NEW YORK

First published 2019
by Routledge
2 Park Square, Milton Park, Abingdon, Oxon OX14 4RN

and by Routledge
605 Third Avenue, New York, NY 10017

First issued in paperback 2020

Routledge is an imprint of the Taylor & Francis Group, an informa business

British Library Cataloguing-in-Publication Data
A catalogue record for this book is available from the British Library

Library of Congress Cataloging-in-Publication Data
Names: Metawa, Noura, editor. | Elhoseny, Mohamed, editor. | Hassanien, Aboul Ella, editor.
Title: Expert systems in finance : smart financial applications in big data environments / edited by Noura Metawa, Mohamed Elhoseny, Aboul Ella Hassanien, M. Kabir Hassan.
Description: 1 Edition. | New York : Routledge, 2019. | Series: Banking, money and international finance | Includes index.
Identifiers: LCCN 2019002509 | ISBN 9780367109523 (hardback)
Subjects: LCSH: Finance—Data processing. | Electronic commerce. | Expert systems (Computer science)
Classification: LCC HG173 .E974 2019 | DDC 332.0285—dc23
LC record available at https://lccn.loc.gov/2019002509

ISBN 13: 978-0-367-72901-1 (pbk)
ISBN 13: 978-0-367-10952-3 (hbk)

Typeset in Times New Roman
by Apex CoVantage, LLC

Contents

Figures

Tables

Contributors

Khalid Abouloula
Department of Mathematics, Informatics and Management, Polydisciplinary
    Faculty of Ouarzazate, Ibn Zohr University, Agadir, Morocco
khabouloula@gmail.com

Mazin A. M. Al Janabi, PhD, SNI II
Full Professor of Finance and Banking and Financial Engineering,
Tecnologico de Monterrey, EGADE Business School, Santa Fe Campus, Mexico
    City, Mexico
mazin.aljanabi@itesm.mx and mazinaljanabi@gmail.com

Dr. Saeed Q. Al-Khalidi Al-Maliki
College of Business, King Khalid University, KSA
salkhalidi@kku.edu.sa

Erfan Babaee Tirkolaee
Department of Industrial Engineering, Mazandaran University of Science and
    Technology, Babol, Iran
E.Babaee@in.iut.ac.ir

Mirjana Pejić Bach
University of Zagreb, Faculty of Economics and Business, Zagreb, Croatia
mpejic@efzg.hr

P. Bangphan
Department of Industrial Engineering, Faculty of Engineering, Rajamangala
    University of Technology Lanna, Thailand
foundry18@yahoo.com

S. Bangphan
Department of Industrial Engineering, Faculty of Engineering, Rajamangala
    University of Technology Lanna, Chiang Mai, Thailand
pong_pang49@yahoo.com

K. Basaid
Laboratory of Mechanic Process Energy and Environment, National School of Applied Sciences, Ibn Zohr University, Agadir, Morocco
Khadija.basaid@gmail.com

R. Bouharroud
Integrated Crop Production Unit, Institut National de la Recherche Agronomique d'Agadir, Agadir, Morocco
bouharroud@yahoo.fr

Jennifer Brodmann
Assistant Professor of Finance, California State University, Dominguez Hills, Carson, CA, USA
jenniferlbrodmann@gmail.com

Esther Castro
Assistant Professor of Finance, Department of Economics and Finance, UHD College of Business, Houston, TX, USA
Email: castroe@uhd.edu

B. Chebli
Laboratory of Mechanic Process Energy and Environment, National School of Applied Sciences, Ibn Zohr University, Agadir, Morocco
b.chebli@uiz.ac.ma

J. N. Furze
Laboratory of Plant Biotechnology, Faculty of Sciences, Ibn Zohr University, Agadir, Morocco
james.n.furze@gmail.com

M. Kabir Hassan
Professor, University of New Orleans, LA, USA
mhassan@uno.edu

Ali Asghar Rahmani Hosseinabadi
Young Researchers and Elite Club, Ayatollah Amoli Branch, Islamic Azad University, Amol, Iran
A.R.Hosseinabadi@iaubeh.ac.ir

Makeen Huda
Candidate and Teaching Associate, Department of Economics and Finance, University of New Orleans, Lakeshore Drive, New Orleans, LA
mhuda1@my.uno.edu

Prayugo Khoir
Student at STMIK Pringsewu, Lampung, Indonesia

Salah-ddine Krit
Department of Mathematics, Informatics and Management, Polydisciplinary Faculty of Ouarzazate, Ibn Zohr University, Agadir, Morocco
salahddine.krit@gmail.com

Xiaolong Liu
College of Computer and Information Sciences, Fujian Agriculture and Forestry University, Fuzhou, China
xlliu@fafu.edu.cn

Andino Maseleno
Sekolah Tinggi Manajemen Informatika dan Komputer (STMIK) Pringsewu, Lampung, Indonesia
andimaseleno@gmail.com

E. H. Mayad
Laboratory of Mechanic Process Energy and Environment, National School of Applied Sciences, and Laboratory of Plant Biotechnology, Faculty of Sciences, Ibn Zohr University, Agadir, Morocco
elhassan.mayad@gmail.com

Muhammad Muslihudin
Lecturer at STMIK Pringsewu, Lampung, Indonesia

Ali Ou-Yassine
Department of Mathematics, Informatics and Management, Polydisciplinary Faculty of Ouarzazate, Ibn Zohr University, Agadir, Morocco
a.ouyassine@yahoo.com

S. Phanphet
Department of Industrial Engineering at Rajamangala University of Technology Lanna, Chiang Mai, Thailand
suwattwong@gmail.com

Saravanan Ramalingam
Department of Computer Science, Pondicherry University, Puducherry, India
r.saravanan26@gmail.com

K. Shankar
School of Computing, Kalasalingam Academy of Research and Education, Krishnankoil, India
shankarcrypto@gmail.com

Pothula Sujatha
Department of Computer Science, Pondicherry University, Puducherry, India
spothula@gmail.com

Phuvadon Wuthisatian
Assistant Professor of Finance, Hastings College, Hastings, NE, USA

1 Theoretical and practical foundations of liquidity-adjusted value-at-risk (LVaR)

Optimization algorithms for portfolio selection and management

Mazin A. M. Al Janabi

Introduction

In general terms, trading risks are usually associated with possible losses in financial markets arising from changes in the price of equities, currencies and interest rates, but there is also risk associated with negative changes in prices of commodities, such as gold and crude oil prices. The measurement and forecasting of financial risks has greatly evolved in the last three decades, from a modest indicator of market value through more multifaceted internal models such as scenario analysis to contemporary stress-testing and value-at-risk (VaR) techniques. Present-day internal regulations for financial markets and institutions formulate some of the market risk management requirements in terms of percentiles of loss distributions; as such, VaR is fundamentally an upper percentile of the loss distribution function.

As a result, it is not surprising to observe that the last three decades have witnessed the growth in academic literature relating alternative modeling methods and suggesting new models and techniques for VaR estimations in an effort to enhance those already in use (Abad et al., 2013). Nevertheless, one of the other categories of risks that still requires further development (within the VaR framework) and incorporation into the structured portfolio optimization and selection process is liquidity trading risk.[1]

Beyond doubt, the recent financial crisis has emphasized the necessity of an adequate assessment of liquidity risk in financial trading portfolios and the evaluation of its impact on the optimization and performance of structured trading portfolios, subject to imposing several financially meaningful operational constraints, under adverse market circumstances and within big data environments. This is the main focus of this research chapter.

The notion of portfolio optimization has received substantial attention in the literature and was pioneered by Markowitz (1952) in the mean-variance (MV) portfolio theory. Indeed, the main focus of the mean-variance portfolio theory is on finding the optimum portfolio when an investor is concerned with the risk/return distributions over a single period. In this framework, risk is defined in terms of the standard deviation of each asset, which implies that the probability of negative returns, as the probability of positive returns, is weighted in the same way by the fund manager. As a result, the solution to the Markowitz theoretical models revolves around the portfolio weights, or the percentage of assets allocated to be invested in each security. However, it is now evident in the academic literature that this approach has several shortcomings, as optimized portfolios normally do not perform as well in practice as in theory (see, for instance, Al Janabi, 2012, 2013 and 2014; Jobson and Korkie, 1981; Jorion, 1991; Michaud, 1989; Fabozzi et al., 2006).[2]

The literature on the application of VaR estimation and forecasting for portfolio optimization and management is rather large and continues to grow, as researchers are continuously attempting to compare alternative optimization algorithms and modeling approaches and proposing new methods and techniques. More recently, focus has shifted from measuring general volatility to tail risk with the popularity of VaR. This simple and intuitive measure is widely used in practice but fails the basic mathematical requirements of rationality (Kaplanski and Kroll, 2002) and coherence (Garcia et al., 2007; Artzner et al., 1999), implying the VaR of a portfolio can be greater than that of the individual assets. To this end, conditional-VaR (CVaR, henceforth), which measures the average loss conditional on returns dropping below the VaR level, is widely documented to be superior to VaR measure. It incorporates more information (Agarwal and Naik, 2004), satisfies all the desirable properties for a coherent risk measure (Rockafellar and Uryasev, 2002; Acerbi and Tasche, 2002) and is much easier to use in optimization techniques (Rockafellar and Uryasev, 2000 and 2002).

Other studies examined portfolio optimization based on variance, VaR and CVaR and under normality assumption (see, for instance, Garcia et al., 2007; Artzner et al., 1999; Ho et al., 2008; Rockafellar and Uryasev, 2002), and found that the resulting risk exposures are equivalent under the three different optimization procedures. It is, however, well-known that asset returns exhibit negative skewness and positive excess kurtosis (evidence was first provided on this by Mandelbrot (1963) but is still a topic of investigation; see Poon and Granger (2005) for an excellent review), with the three optimization procedures generating very different results under these conditions (Gaivoronski and Pflug, 2005; Agarwal and Naik, 2004).

Indeed, methodologies for the assessment of market risk have been well established and standardized in the academic world as well as the financial markets. On the other hand, liquidity risk has gained less devotion from researchers, possibly because it is less substantial in developed countries where most of the market risk methodologies and techniques were devised (Al Janabi, 2013). Nevertheless, the recent rapid expansion of emerging markets' trading activities as well as the recurring turbulence in those markets, in the wake of the outcomes of the most recent financial crunch, has driven liquidity risk and its application to the portfolio optimization enigma to the vanguard of market risk research and development (Al Janabi, 2013).

In fact, some relevant studies have tackled the issues of liquidity risk but not necessarily within the context of optimization of structured trading portfolios.

Their main focus, in fact, was on modeling only transaction costs (that is, the widening of the bid-ask spread), however the effects of adverse market price impact have not been studied rigorously, although Al Janabi (2012, 2013 and 2014) is an exception. For the sake of brevity, we discuss below some of the suggested modeling techniques for liquidity risk, detailed as follows. Within the VaR technique, Jarrow and Subramanian (1997) provide a market impact model of liquidity by considering the optimal liquidation of an investment portfolio over a fixed horizon. On the other hand, Bangia et al. (2002) approach the liquidity risk from another angle and provide a model of VaR adjusted for what they call exogenous liquidity, defined as common to all market players and unaffected by the actions of any one participant. It comprises such execution costs as order processing costs and adverse selection costs resulting in a given bid-ask spread faced by investors in the market. In a different vein, Almgren and Chriss (1999) present a concrete framework for deriving the optimal execution strategy using a mean-variance approach and show a specific calculation method.[3]

In fact, despite some research into liquidity risk measurement, the financial services industry still finds it difficult to quantify and predict measurable attributes of market liquidity risk and its application to the problem of portfolio optimization. As indicated earlier, liquidity risk is challenging to assess and forecast since it depends on so many interlinked factors, and the latest financial crunch has highlighted the growing prominence of liquidity risk as an integral part of risk management process and the need for more efficient assessment techniques of liquidity risk under stress market conditions (Ruozi and Ferrari, 2013). This is where this research chapter comes in, as we strive to clarify the essence of liquidity risk, provide clear definition of the topic and suggest methods for its assessment and forecasting at the level of multi-asset portfolios and its use for the optimization of structured portfolios subject to imposing several financially meaningful operational constraints under adverse market circumstances.

With this backdrop and to address the above shortcomings, this chapter reviews and examines robust techniques for the optimization of financial trading portfolios and for forecasting and assessing liquidity risk within a multivariate technique (recently expanded by Al Janabi (2012, 2014).[4] Given the fact that literature on modern portfolio management has been relatively inconclusive and provided mixed results, the aim of this chapter is to examine realistic optimization algorithms that can produce better multi-asset allocation under event (crisis) market circumstances and construct realistic structured portfolios, subject to applying meaningful operational and financial constraints, predominantly in light of the shocks of the recent financial crunch.

To this end, the main contribution of this chapter is to develop an optimization algorithm to improve the asset allocation process of multi-asset portfolios by combining LVaR with a multivariate dependence modeling technique. Our proposed methodology differs in important ways from existing published literature pertaining to the application of advanced assessment techniques to liquidity risk analysis and optimization of trading portfolios.

As such, our proposed approach is a robust enhancement of the traditional Markowitz (1952) MV approach, where the variance risk measure is replaced by multivariate LVaR optimization algorithms. The task is achieved by minimizing the LVaR while requiring a minimum expected return subject to imposing several financially meaningful operational constraints under event (adverse) market circumstances. As such, the focus in this work is on the forecast of risk measure rather than on expected returns for two reasons: first, several studies have analyzed the forecasts of expected returns in the context of MV optimization (see, for instance, Best and Grauer, 1991). The common opinion is that expected returns are not easy to forecast and that the optimization process is very sensitive to these variations. Second, there exists a general notion that LVaR, in a wide sense, is simpler to assess than expected returns from historical data (Al Janabi, 2013 and 2014). To that end, a comparison between the classical Markowitz (1952) MV efficient portfolio and investable LVaR portfolio is obtained by simulating different expected returns under event market conditions. Lastly, we show the usefulness of our optimization algorithm for portfolio selection by considering selected diversified portfolios of emerging Gulf Cooperation Council (GCC) stock market indices and two benchmark indices. To the present author's knowledge, none of the studies in the literature has done this type of empirical testing by comparing daily LVaR risk measures and investable portfolios with the conventional Markowitz (1952) MV efficient portfolio technique. Finally, the empirical testing results are interesting in terms of theory as well as practical application and provide an incentive for further research in the areas of LVaR optimization algorithms, asset allocation and portfolio selection of equity investable portfolios. Moreover, the different obtained asset allocation and optimized investable portfolio case analysis studies are widely applicable to any equity portfolio management end user, providing potential applications to practitioners and research ideas to academics.

The chapter proceeds as follows. The next section outlines the theoretical foundations and modeling parameters for portfolio selection using LVaR optimization algorithms. This is followed by full discussion of the optimization algorithm for portfolio selection and validation of the optimization method and research design for portfolio selection. This section also discusses and analyses the various parameters required for the optimization engine and the construction of efficient and coherent portfolios for long and short sales trading positions and long-only portfolio. The final section includes concluding remarks along with possible practical applications for portfolio selection and asset management. Appendix A includes detailed discussion on the optimization algorithm for portfolio selection and research design for computer programming purposes, and highlights the required input parameters for robust quantitative risk analysis and optimization algorithms for investable portfolio selection.

Theoretical foundations and modeling parameters for portfolio selection

LVaR model for risk evaluation under event market viewpoints

To calculate VaR, the volatility of each risk factor is extracted from a pre-defined historical observation period and can be estimated using the GARCH-M (1,1) model under the assumptions of adverse market settings. The potential effect of each component of the portfolio on the overall portfolio value is then worked out. These effects are then aggregated across the whole portfolio using the correlations between the risk factors (which are, again, extracted from the historical observation period) to give the overall VaR of the portfolio with a given confidence level. As such, for a single trading position the absolute value of VaR can be defined in monetary terms as follows:

$$VaR_i =|(\mu_i - \alpha * \sigma_i)(Asset_i * Fx_i)| \qquad (1.1)$$

where μ_i is the expected return of asset i, α is the confidence level (or in other words, the standard normal variant at confidence level α) and σ_i is the conditional volatility of the return of the security that constitutes the single position and can be estimated using a GARCH-M (1,1) model.[5] While the term $Asset_i$ indicates the mark-to-market monetary value of asset i, Fx_i denotes the unit foreign exchange rate of asset i. If the expected return of the asset, μ_i, is very small or close to zero, then equation 1.1 can be reduced to:[6]

$$VaR_i =|\alpha * \sigma_i * Asset_i * Fx_i| \qquad (1.2)$$

Indeed, equation 1.2 includes some simplifying assumptions, yet it is routinely used by researchers and practitioners in the financial markets for the estimation of VaR for a single trading position.

Trading risk in the presence of multiple risk factors is determined by the combined effect of individual risks. The extent of the total risk is determined not only by the magnitudes of the individual risks but also by their correlations. Portfolio effects are crucial in risk management not only for large, diversified portfolios but also for individual instruments and depend on several risk factors. For multiple assets or portfolio of assets, VaR is a function of each individual security's risk and the Pearson correlation factor $[\rho_{i,j}]$ between the returns on the individual securities as follows:

$$VaR_P = \sqrt{\sum_{i=1}^{n} \sum_{j=1}^{n} VaR_i \, VaR_j \rho_{i,j}} = \sqrt{[VaR]^T [\rho] [VaR]} \qquad (1.3)$$

This formula is a general one for the calculation of VaR for any portfolio regardless of the number of securities. It should be noted that the second term of equation 1.3 is rewritten in terms of matrix algebra – a useful form to avoid mathematical complexity as more and more securities are added. This approach can simplify the programming process and permits easy incorporation of short sales positions in market risk management process. This means that in order to calculate VaR (of a portfolio of any number of securities), we need to create first a transpose vector $[VaR]^T$ of individual VaR positions, a $(1 \times n)$ vector, and hence the superscript T indicates transpose of the vector:

$$[VaR]^T = [VaR_1 \quad VaR_2 \quad .... \quad VaR_n] \qquad (1.3a)$$

Second, a vector $[VaR]$ of individual VaR positions, explicitly n rows and one column ($n \times 1$) vector, such as:

$$[VaR] = \begin{bmatrix} VaR_1 \\ VaR_2 \\ ..... \\ VaR_n \end{bmatrix} \tag{1.3b}$$

Finally, a matrix $[\rho]$ of all correlation factors (ρ), an ($n \times n$) matrix in the following form:

$$[\rho] = \begin{bmatrix} 1 & \rho_{1,2} & \rho_{1,3} & .... & \rho_{1,n} \\ \rho_{2,1} & 1 & \rho_{2,3} & .... & \rho_{2,n} \\ \rho_{3,1} & \rho_{3,2} & 1 & .... & \rho_{3,n} \\ ...... & .... & ...... & .... & .... \\ \rho_{n,1} & \rho_{n,2} & \rho_{n,3} & .... & 1 \end{bmatrix} \tag{1.3c}$$

Consequently, as one multiplies the two vectors and the correlation matrix and then takes the square root of the result, one ends up with the VaR_p of any portfolio with any n-number of securities. This simple number summarizes the portfolio's exposure to market risk. Portfolio managers can then decide whether they feel comfortable with this level of risk. If the answer is no, then the process that led to the estimation of VaR can be used to decide where to reduce redundant risk. For instance, the riskiest assets can be sold, or one can use derivative securities such as futures and options contracts on these particular assets to hedge the undesirable risk.

In effect, the VaR method is only one approach of measuring market risk and is mainly concerned with maximum expected losses under normal market conditions. It is not an absolute measure, as the actual amount of loss may be greater than the given VaR amounts under severe circumstances. In extreme situations, VaR models do not function very well. As a result, for prudent risk management and as an extra management tool, firms should augment VaR analysis with stress-testing and scenario procedures. From a risk management perspective, however, it is desirable to have an estimate for what potential losses could be under severely adverse conditions where statistical tools do not apply. As such, stress-testing usually takes the form of subjectively specifying scenarios of interest to assess changes in the value of the portfolio, and it can involve examining the effect of past large market moves on today's portfolio.

Modeling parameters for incorporating asset liquidity risk into LVaR models

Liquidity is a key risk factor that until lately has not been appropriately dealt with by risk models. Illiquid trading positions can add considerably to losses and can give negative signals to traders due to the higher expected returns they entail. The concept of liquidity trading risk is immensely important for using VaR accurately, and recent upheavals in financial markets confirm the need for laborious treatment and assimilation of liquidity trading risk into VaR models.

The simplest way to account for liquidity trading risk is to extend the holding period of illiquid positions to reflect a suitable liquidation period. An adjustment can be made by adding a multiplier to the VaR measure of each trading asset type, which at the end depends on the liquidity of each individual security. Nonetheless, the weakness of this method is that it allows for subjective assessment of the liquidation period. Furthermore, the typical assumption of a one-day horizon (or any inflexible time horizon) within the VaR framework neglects any calculation of trading risk related to the liquidity effect (that is, when and whether a trading position can be sold out and at what price). A broad VaR model should incorporate a liquidity premium (or liquidity risk factor). This can be worked out by formulating a method by which one can unwind a position, not at some ad hoc rate but at the rate that market conditions are optimal, so that one can effectively set a risk value for the liquidity effects. In general, this will significantly raise the VaR, or the amount of economic capital to support the trading position.

In fact, if returns are independent and they can have any elliptical multivariate distribution, then it is possible to convert the VaR horizon parameter from daily to any $t-$ day horizon. The variance of a $t-$ day return should be t times the variance of a $1-$ day return or $\sigma^2 = f(t)$. Thus, in terms of standard deviation (or volatility), $\sigma = f(\sqrt{t}\,)$ and the daily or overnight VaR number [*VaR* $(1-day)$], it is possible to determine the liquidity-adjusted VaR (LVaR) for any *t-day* horizon as:

$$LVaR\ (t-day) = VaR\ (1-day)\sqrt{t} \tag{1.4}$$

The preceding formula was proposed and used by J.P. Morgan in their earlier RiskMetrics method (Morgan Guaranty Trust Company, 1994). This methodology implicitly assumes that liquidation occurs in one block sale at the end of the holding period and that there is one holding period for all assets, regardless of their inherent trading liquidity structure. Unfortunately, the latter approach does not consider real-life trading situations, where traders can liquidate (or rebalance) small portions of their trading portfolios on a daily basis. The assumption of a given holding period for orderly liquidation inevitably implies that asset liquidation occurs during the holding period. Accordingly, scaling the holding period to account for orderly liquidation can be justified if one allows the assets to be liquidated throughout the holding period.

In what follows, we examine a re-engineered approach for computing a closed-form LVaR with explicit treatment of liquidity trading risk and coherent assessment of investable portfolios.[7] The key methodological contribution is a different liquidity-scaling factor than the traditional root-*t* multiplier. The proposed model and liquidity-scaling factor is more realistic and less conservative than the conventional root-*t* multiplier. In essence, the suggested multiplier (add-on) is a function of a predetermined liquidity threshold, defined as the maximum position which can be unwound without disturbing market prices during one trading day. The essence of the model relies on the assumption of a stochastic stationary process and some rules of thumb, which can be of crucial value for more accurate overall trading risk assessment during market stress periods when liquidity dries up in consequence of a financial critical situation. In addition, the re-engineered model is quite simple to implement even by very large financial institutions with multiple assets and risk factors. To this end, a practical framework of a methodology (within a simplified mathematical approach) is proposed below with the purpose of incorporating and calculating illiquid assets' horizon LVaR, detailed along these lines:[8]

> The market risk of an illiquid asset trading position is larger than the risk of an otherwise identical liquid position. This is because unwinding the illiquid position takes longer than unwinding the liquid position, and as a result, the illiquid position is more exposed to the volatility of the market for a longer period of time. In this approach, an asset trading position will be thought of illiquid if its size surpasses a certain liquidity threshold. The threshold (which is determined by traders for different assets and/or financial markets) is defined as the maximum position which can be unwound, without disrupting market prices, in normal market conditions and during one trading day. Consequently, the size of the asset trading position relative to the threshold plays an important role in determining the number of days that are required to close the entire position. This effect can be translated into a liquidity increment (or an additional liquidity risk factor) that can be incorporated into VaR analysis. If for instance, the par value of an asset position is $250,000 and the liquidity threshold is $50,000, then it will take four days to sell out the entire trading position. Therefore, the initial position will be exposed to market variation for one day, and the rest of the position (that is, $200,000) is subject to market variation for an additional three days. If it assumed that daily changes of market values follow a stationary stochastic process, the risk exposure due to illiquidity effects is given by the following illustration, detailed as follows:

>> In order to take into account the full illiquidity of assets (that is, the required unwinding period to liquidate an asset), we define the following:

>> t = number of liquidation days (t days to liquidate the entire asset fully)
>> σ_{adj}^2 = variance of the illiquid asset trading position
>> σ_{adj} = liquidity risk factor or standard deviation of the illiquid asset trading position.

The proposed approach assumes that the trading position is closed out linearly over t days, and hence it uses the logical assumption that the losses due to illiquid trading positions over t days are the sum of losses over the individual trading days. Moreover, we can assume with reasonable accuracy that asset returns and losses due to illiquid trading positions are independent and identically distributed (IID) and serially uncorrelated day to day along the liquidation horizon, and that the variance of losses due to liquidity risk over t days is the sum of the variance (σ_i^2 for all $i = 1, 2, \ldots, t$) of losses on the individual days, thus:

$$\sigma_{adj}^2 = \left(\sigma_1^2 + \sigma_2^2 + \sigma_3^2 + \Lambda + \sigma_{t-2}^2 + \sigma_{t-1}^2 + \sigma_t^2 \right) \tag{1.5}$$

In fact, the square root-t approach (equation 1.4) is a simplified special case of equation 1.5 under the assumption that the daily variances of losses throughout the holding period are all the same as the first-day variance, σ_1^2, thus $\sigma_{adj}^2 = \left(\sigma_1^2 + \sigma_1^2 + \sigma_1^2 + \Lambda + \sigma_1^2 \right) = t\,\sigma_1^2$. As discussed above, the square root-t equation overestimates asset liquidity risk since it does not consider that traders can liquidate small portions of their trading portfolios on a daily basis and then the whole trading position can be sold completely on the last trading day. However, this would be an overstatement of VaR, and the true VaR has to be between the 1-day VaR and 1-day VaR $\sqrt{t}$.

Indeed, in real financial market operations, liquidation occurs during the holding period, and thus scaling the holding period to account for orderly liquidation can be justified if one allows the assets to be liquidated throughout the holding period. As such, for this special linear liquidation case, and under the assumption that the variance of losses of the first trading day decreases linearly each day (as a function of t), we can derive from equation 1.5 the following:

$$\sigma_{adj}^2 = \begin{pmatrix} (\frac{t}{t})^2 \sigma_1^2 + (\frac{t-1}{t})^2 \sigma_1^2 + (\frac{t-2}{t})^2 \sigma_1^2 + \Lambda \\ + (\frac{3}{t})^2 \sigma_1^2 + (\frac{2}{t})^2 \sigma_1^2 + (\frac{1}{t})^2 \sigma_1^2 \end{pmatrix} \tag{1.6}$$

In this manner, if the asset position is liquidated in equal parts at the end of each trading day, the trader faces a 1-day holding period on the entire position, a 2-day holding period on a fraction $(t - 1)/t$ of the position, a 3-day holding period on a fraction $(t - 2)/t$ of the position and so forth. Evidently, the additional liquidity risk factor depends only on the number of days needed to sell an illiquid trading position linearly. In the general case of t days, the variance of the liquidity risk factor is given by the following mathematical functional expression of t:

$$\sigma_{adj}^2 = \sigma_1^2 \left((\frac{t}{t})^2 + (\frac{t-1}{t})^2 + (\frac{t-2}{t})^2 + \Lambda + (\frac{3}{t})^2 + (\frac{2}{t})^2 + (\frac{1}{t})^2 \right) \tag{1.7}$$

To calculate the sum of the squares, it is convenient to use a shortcut approach. From mathematical finite series, the following relationship can be obtained:

$$(t)^2 + (t-1)^2 + (t-2)^2 + \Lambda + (3)^2 + (2)^2 + (1)^2 = \frac{t(t+1)(2t+1)}{6} \tag{1.8}$$

Hence, after substituting equation 1.8 into equation 1.7, the following can be achieved:

$$\sigma_{adj}^{2} = \sigma_1^2 \frac{1}{t^2}\left[\left\{(t)^2 + (t-1)^2 + (t-2)^2 + \Lambda + (3)^2 + (2)^2 + (1)^2\right\}\right]$$

$$\text{or } \sigma_{adj}^{2} = \sigma_1^2 \left(\frac{(2t+1)\,(t+1)}{6t}\right) \tag{1.9}$$

Accordingly, from equation 1.9 the liquidity risk factor can be expressed in terms of volatility (or standard deviation) as:

$$\sigma_{adj} = \sigma_1 \left\{ \sqrt{\frac{1}{t^2}[(t)^2 + (t-1)^2 + (t-2)^2 + \cdots + (3)^2 + (2)^2 + (1)^2}\right\}$$

$$\text{or } \sigma_{adj} = \sigma_1 \left\{ \sqrt{\frac{(2t+1)\,(t+1)}{6t}}\right\} \tag{1.10}$$

The result of equation 1.10 is of course a function of time and not the square root of time, as employed by some financial market participants based on the Risk-Metrics methodologies. The above approach can also be used to calculate LVaR for any time horizon. Likewise, in order to perform the calculation of LVaR under illiquid market conditions, it is possible to use the liquidity factor of equation 1.10 and define the following:[9]

$$LVaR_{adj} = VaR \sqrt{\frac{(2t+1)\,(t+1)}{6t}} \tag{1.11}$$

where VaR = value at risk under liquid market conditions and $LVaR_{adj}$ = value at risk under illiquid market conditions. The latter equation indicates that $LVaR_{adj} >$ VaR, and for the special case when the number of days to liquidate the entire assets is one trading day, then $LVaR_{adj} = VaR$. Consequently, the difference between $LVaR_{adj}$ and VaR should be equal to the residual market risk due to the illiquidity of any asset under illiquid market conditions. In fact, the number of liquidation days (t) necessary to liquidate the entire asset fully is related to the choice of the liquidity threshold; however, the size of this threshold is likely to change under severe market conditions. Indeed, the choice of the liquidation horizon can be estimated from the total trading position size and the daily trading volume that can be unwound into the market without significantly disrupting asset market prices; in actual practice it is generally estimated as:

$$t = Total\ Trading\ Position\ Size\ of\ Asset_i\ /$$
$$Daily\ Trading\ Volume\ of\ Asset_i \tag{1.12}$$

In practice, the daily trading volume of any trading asset is estimated as the average volume over some period of time, generally a month of trading activities. In effect, the daily trading volume of assets can be regarded as the average daily volume or the volume that can be unwound in a severe event (crisis) period. The trading volume in an event period can be roughly approximated as the average daily trading volume less a number of standard deviations. Although this alternative approach is quite simple, it is still relatively objective. Moreover, it is reasonably easy to gather the required data to perform the necessary liquidation scenarios.

In essence, the above liquidity scaling factor (or multiplier) is more realistic and less conservative than the conventional root-*t* multiplier and can aid financial entities in allocating reasonable and liquidity market-driven regulatory and economic capital requirements. Furthermore, the above mathematical formulas can be applied for the calculation of LVaR for every asset trading position and for the entire portfolio of assets. In order to calculate the LVaR for the full trading portfolio under illiquid market conditions ($LVaR_{Padj}$), the above mathematical formulation can be extended, with the aid of equation (1.3), into a matrix algebra form to yield the following:

$$LVaR_{P_{adj}} = \sqrt{\sum_{i=1}^{n}\sum_{j=1}^{n} LVaR_{i_{adj}} LVaR_{j_{adj}} \rho_{i,j}}$$

$$= \sqrt{\left[LVaR_{adj} \right]^{T} \left[\rho \right] \left[LVaR_{adj} \right]}$$ (1.13)

The elements of the vectors of equation 1.13 (i.e. the absolute value of $LVaR_{iadj}$) for each trading asset can now be calculated with the aid of equations 1.1, 1.2 and 1.11 in this manner:

$$LVaR_{i adj} = |(m_i - a * s_i) Asset_i * Fx_i \sqrt{\frac{(2t_i + 1)(t_i + 1)}{6t_i}}|$$ (1.14)

On the other hand, for the special case when μ_i is small or close to zero, we can have:

$$LVaR_{i adj} = |a * s_i * Asset_i * Fx_i \sqrt{\frac{(2t_i + 1)(t_i + 1)}{6t_i}}|$$ (1.14a)

Now we can define the ultimate two vectors $[LVaR_{adj}]^{T}$ and $[LVaR_{adj}]$ as follows:

$$\left[LVaR_{adj} \right]^{T} = \left[LVaR_{1_{adj}} \quad LVaR_{2_{adj}} \quad .... \quad LVaR_{n_{adj}} \right]$$ (1.15)

$$\left[LVaR_{adj} \right] = \begin{bmatrix} LVaR_{1_{adj}} \\ LVaR_{2_{adj}} \\ .... \\ LVaR_{n_{adj}} \end{bmatrix}$$ (1.16)

The above mathematical structure (in the form of two vectors and a matrix, $[LVaR_{adj}]^T$, $[LVaR_{adj}]$, and $[\rho]$) can facilitate the computer programming process so that the portfolio manager can specify different liquidation days for the whole portfolio and/or for each individual trading security according to the necessary number of days to liquidate the entire asset fully. The latter can be achieved by specifying an overall benchmark liquidation horizon to liquidate the entire contents of the portfolio fully. The number of days required to liquidate trading asset positions (of course, depending on the type of asset) can be obtained from the various publications in financial markets and can be compared with the assessments of individual traders of each trading unit. As a result, it is possible to create simple statistics of the asset volume that can be liquidated and the necessary time horizon to unwind the whole volume.

Optimization algorithm for portfolio selection

Set against this background, asset return can be defined as $R_{i,t} = ln(P_{i,t}) - ln(P_{i,t-1})$, where $R_{i,t}$ is the daily return of asset i, ln is the natural logarithm, $P_{i,t}$ is the current price level of asset i, and $P_{i,t-1}$ is the previous day price level. Furthermore, for this research study we can, for instance, choose a confidence interval of 95% (or 97.5% with "one-tailed" loss side) and several liquidation time horizons to compute LVaR. In the process of analyzing the data, first the daily log returns of the assets should be calculated. These daily returns are essential inputs for the calculation of conditional volatilities, expected returns and correlation matrices under adverse market outlooks.

To examine the relationship between asset expected returns and volatility, we can implement a conditional volatility approach to determine the risk parameters that are needed for the LVaR's engine and thereafter for the estimation of daily asset market liquidity risk exposure and investable portfolio requirements. Indeed, the time-varying pattern of asset volatility has been widely recognized and modeled as a conditional variance within the GARCH framework, as originally developed by Engle (1982, 1995). Engle (1982) introduced a likelihood ratio test to ARCH effects and a maximum likelihood method to estimate the parameters in the ARCH model. This approach was generalized by Bollerslev (1986) and Engle and Kroner (1995). The following generalized autoregressive conditional heteroskedasticity in mean, GARCH-M (1,1) model, is used for the estimation of expected return and conditional volatility for each of the time-series variables:[10]

$$R_{it} = a_i + b_i \sigma_{it} + \varepsilon_{it}, \tag{1.17}$$

$$\sigma_{it}^2 = c_i + \beta_{i1} \sigma_{it-1}^2 + \beta_{i2} \varepsilon_{it-1}^2, \tag{1.18}$$

where R_{it} is the continuous compounding return of time series i, σ_{it} is the conditional standard deviation as a measure of volatility, and ε_{it} is the error term return for time series i. The denotations a_i, b_i, c_i, β_{i1} and β_{i2} represent parameters to be estimated. The parameters representing variance are assumed to undertake a positive value.

Furthermore, three matrices of correlations can possibly be used for the assessment of LVaR, namely $\rho = 1$, 0 and empirical correlations. The objectives here are to establish the necessary quantitative infrastructures for advanced portfolio selection and management. Indeed, the assembled correlation matrices are essential parameters along with conditional volatility vectors for the calculations of LVaR and stress-testing and LVaR of efficient and investable portfolios. As a result, these matrices and vectors can be integrated into the LVaR's engine to estimate efficient and investable portfolios under the notion of different unwinding horizon periods and for a combination of varied long- and short-sales asset trading positions (that is, by disallowing both long-only positions and borrowing constraints).

Optimization parameters and constraints for portfolio selection using LVaR modeling algorithm

In this chapter, we examine a model for optimizing portfolio risk-return with LVaR constraints using realistic operational and financial scenarios. Essentially, our proposed approach is a robust enhancement of the classic Markowitz (1952) mean-variance approach, where the original risk measure (variance) is replaced by LVaR algorithms. The task is attained here by minimizing LVaR while requiring a minimum expected return subject to imposing several financially meaningful operational constraints under adverse market circumstances. Thus by considering different expected returns, we can generate an optimal LVaR frontier under adverse outlooks. Alternatively, we can also maximize expected returns while not allowing for large risks. For the purposes of this study, the optimization problem can be formulated as follows:

It is possible to compute from equation 1.13 the minimum portfolio LVaR by solving for the following quadratic programming objective function:

$$\text{Minimize}: LVaR_{P_{adj}} = \sqrt{\sum_{i=1}^{n} \sum_{j=1}^{n} LVaR_{i_{adj}} \, LVaR_{j_{adj}} \, \rho_{i,j}}$$

$$= \sqrt{\left[LVaR_{adj}\right]^{T} \left[\rho\right] \left[LVaR_{adj}\right]} \tag{1.19}$$

The above objective function can be minimized subject to the following operational and financial budget constraints (or boundary conditions) as specified by the portfolio manager:

$$\sum_{i=1}^{n} R_i x_i = R_P \; ; \; l_i \leq x_i \leq u_i \quad i = 1, 2, \ldots, n \tag{1.20}$$

$$\sum_{i=1}^{n} x_i = 1.0 \; ; \; l_i \leq x_i \leq u_i \quad i = 1, 2, \ldots, n \tag{1.21}$$

$$\sum_{i=1}^{n} V_i = V_P \quad i = 1, 2, \ldots, n \tag{1.22}$$

$$\left[LHF\right] \geq 1.0 \; ; \; \forall_i \quad i = 1, 2, \ldots, n \tag{1.23}$$

Here R_p and V_p denote the target portfolio mean expected return and total portfolio volume, respectively, and x_i the weight or percentage asset allocation for each asset. The values l_i and u_i, $i = 1, 2, \ldots, n$, denote the lower and upper constraints for the portfolio weights x_i. If we choose $l_i = 0$, $i = 1, 2, \ldots, n$, then we have the situation where no short-sales are allowed. Moreover, $[LHF]$ indicates an $(n-1)$ vector of the individual liquidity horizon of each asset for all $i = 1, 2, \ldots, n$, where LHF_i is defined, with the aid of equation 1.11, for each trading asset in this way:

$$LHF_i = \sqrt{\frac{(2t_i + 1)(t_i + 1)}{6t_i}} \geq 1.0 \; ; \forall_i \quad i = 1, 2, \ldots, n \tag{1.24}$$

Now the portfolio manager can specify different liquidity horizon and correlation factors and calculate the necessary LVaR to sustain the trading operation of the financial entity without subjecting the entity to insolvency matters. The rationality behind imposing the above constraints is to comply with current regulations that enforce capital requirements on investment companies, proportional to their LVaR and economic capital, besides other operational limits (e.g. volume trading limits).

Validation of optimization method and research design for portfolio selection

Against this backdrop, one case where the application of the optimization algorithm with LVaR technique is useful is in its implementation for stock market portfolio selection for asset management firms as a whole. As such, the above developed optimization methodology and algorithms are applied to diverse stock markets selected portfolios of the GCC zone, including:

1   ADSMI Index (United Arab Emirates, Abu Dhabi Stock Market Index)
2   DFM General Index (United Arab Emirates, Dubai Financial Market General Index)
3   KSE General Index (Kuwait, Stock Exchange General Index)
4   BA All Share Index (Bahrain, All Share Stock Market Index)
5   MSM30 Index (Oman, Muscat Stock Market Index)
6   DSM20 Index (Qatar, Doha Stock Market General Index)
7   SE All Share Index (Saudi Arabia, All Share Stock Market Index)
8   Shuaa GCC Index (Shuaa Capital, GCC Stock Markets Benchmark Index)
9   Shuaa Arab Index (Shuaa Capital, Arab Stock Markets Benchmark Index).

For this particular research study we have chosen a confidence interval of 95% (or 97.5% with "one-tailed" loss side) and several liquidation time horizons to compute LVaR. Historical GCC zone stock market databases of more than six years of daily closing index levels, for the period 17 October 2004–22 May 2009 are collected for the sake of implementing this research study as well as for the assessment of market risk management parameters, optimization algorithms and portfolio selection.[11] In fact, the selected time-series datasets fall within the period of the most severe part of the latest subprime financial crunch and represent a typical case within big data environments.

To investigate the statistical and stochastic properties of the data, we have computed the log returns of each series. As such, Table 1.1 illustrates the daily volatility of each of the sample indices under regular and adverse (event) market conditions along with the assessment of systematic risk factors (beta coefficient). Event market volatilities are calculated by implementing an empirical distribution of past returns for all stock market indices' time series and, hence the maximum negative returns (losses), which are witnessed in the historical time series, are selected for this purpose. This approach can aid in overcoming some of the limitations of normality assumption and can provide a better analysis of LVaR and especially under event and illiquid market settings.

In another study, the measurements of skewness and kurtosis are achieved on the sample indices and the results are depicted in Table 1.1. In general, it is perceived that all stock market indices have shown asymmetric behavior (both positive and negative values). Moreover, kurtosis studies have shown similar patterns of abnormality. As a result, the Jarque-Bera (JB) test shows an obvious general deviation from normality and thus rejects the hypothesis that GCC stock markets' time series returns are normally distributed.

For the sake of illustrating the methodological aspects of the optimization algorithm for portfolio selections, we present a real-world operational and optimization process by relaxing the restriction on short-sales of trading assets. As such, Figure 1.1 provides evidence of the empirical MV efficient frontier – under event (adverse) market settings – defined using a 97.5% confidence level. In this case, the efficient portfolio selection is performed by relaxing the short-sale constraint for the different GCC stock markets.

Table 1.1 Stochastic properties of the return series of stock market indices

Stock Market Indices	Volatility (Regular Market)	Volatility (Event Market)	Expected Return	Systematic Risk Factor (Beta)	Skewness	Kurtosis	Jarque-Bera (JB) Test
DFM General Index	1.93%	12.16%	0.12%	0.58	0.01	7.86	955**
ADSMI Index	1.42%	7.08%	0.07%	0.40	0.12	7.26	734**
BA All Share Index	0.59%	3.77%	0.05%	0.06	0.43	10.24	2142**
KSE General Index	0.76%	3.74%	0.09%	0.14	−0.18	8.38	1173**
MSM30 Index	0.84%	8.70%	0.12%	0.10	−0.57	18.40	9617**
DSM20 Index	1.53%	8.07%	0.06%	0.31	−0.11	5.59	273**
SE All Share Index	2.08%	11.03%	0.03%	0.98	−0.97	8.47	1361**
Shuaa GCC Index	1.45%	8.10%	0.06%	1.05	−0.66	14.00	4949**
Shuaa Arab Index	1.28%	7.57%	0.07%	1.00	−0.61	13.79	4758**

Notes: (1) Asterisk ** denotes statistical significance at the 0.01 level. JB test indicates the absence of normality for all stock market indices. (2) Downside risk under event market conditions is simulated as the conditional volatility of the maximum negative daily return (losses).

Source: Designed by the author using in-house developed software.

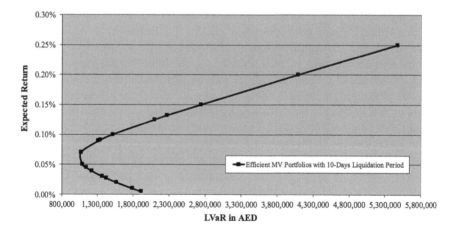

Figure 1.1 Efficient portfolios with Markowitz's mean-variance (MV) technique

This figure demonstrates the empirical LVaR efficient frontier under event market scenarios by relaxing short-sales constraints (i.e. allowing both long- and short-sales trading positions).

Source: Designed by the author using in-house developed software.

On the other hand, efficient portfolios cannot always be attained (e.g. short-sales without realistic lower boundaries on x_j) in the day-to-day, real-world portfolio management operations and, hence, portfolio managers should establish proactive investable portfolios under more realistic and restricted budget constraints, as follows:[12]

- The total trading volume (between long and short-sales equity trading positions) of any investable portfolio is 10 million AED (UAE Dirhams).
- For any investable portfolio, the asset allocation for a long equity trading position varies from 10% to 50%.
- For any investable portfolio, the asset allocation for short-sales equity trading position varies from −10% to −30%.
- All liquidity horizons (closeout periods) for all equities are kept constant and in accordance with the values indicated in Tables 1.2–1.5.
- Volatilities under event market notions are estimated as the maximum historical simulation events with the highest downside risk. These conditional volatilities are kept constant throughout the optimization process and in accordance with the estimated values indicated in Tables 1.2–1.5.

For the same reason, the optimization process is based on the definition of LVaR as the minimum possible loss over a specified time horizon within a given confidence level. The optimization technique solves the problem by finding the market long- and short-sales trading positions that minimize the loss, subject to the fact that all constraints are satisfied within their boundary values. As such,

Table 1.2 Asset allocation and composition of investable market portfolio [1]

Market Index	Liquidation Period (in Days)	Market Value in AED	Asset Allocation per Market
DFM General Index	2.0	−1,500,000	−15%
ADSMI Index	3.0	3,000,000	30%
BA All Share Index	4.0	4,000,000	40%
KSE General Index	3.0	3,000,000	30%
MSM30 Index	4.0	4,000,000	40%
DSM20 Index	3.0	−1,000,000	−10%
SE All Share Index	2.0	−1,500,000	−15%
Correlation Factor	**LVaR (Crisis Market)**	**LVaR/Volume**	**Expected Return**
ρ = Empirical	1,233,762	12.3%	0.085%
ρ = 1	1,195,577	12.0%	**Systematic Risk Factor (Beta)**
ρ = 0	1,334,455	13.3%	−0.040

This table illustrates the asset allocation and composition of investable portfolio [1] using long- and short-sales trading positions under event (crisis) market conditions. The table also depicts LVaR under event market settings with different correlation factors, along with expected return and systematic risk factor (beta coefficient).

Source: Designed by the author using in-house developed software.

Table 1.3 Asset allocation and composition of investable market portfolio [2]

Market Index	liquidation Period (in Days)	Market Value in AED	Asset Allocation per Market
DFM General Index	2.0	−3,000,000	−30%
ADSMI Index	3.0	3,500,000	35%
BA All Share Index	4.0	5,000,000	50%
KSE General Index	3.0	5,000,000	50%
MSM30 Index	4.0	4,500,000	45%
DSM20 Index	3.0	−3,000,000	−30%
SE All Share Index	2.0	−2,000,000	−20%
Correlation Factor	**LVaR (Crisis Market)**	**LVaR/Volume**	**Expected Return**
ρ = Empirical	1,550,118	15.5%	0.086%
ρ = 1	758,616	7.6%	**Systematic Risk Factor (Beta)**
ρ = 0	1,813,254	18.1%	−0.179

This table illustrates the asset allocation and composition of investable portfolio [2] using long- and short-sales trading positions under event (crisis) market conditions. The table also depicts LVaR under event market settings with different correlation factors, along with expected return and systematic risk factor (beta coefficient).

Source: Designed by the author using in-house developed software.

Table 1.4 Asset allocation and composition of investable market portfolio [3]

Market Index	Liquidation Period (in Days)	Market Value in AED	Asset Allocation per Market
DFM General Index	2.0	3,000,000	30%
ADSMI Index	3.0	2,000,000	20%
BA All Share Index	4.0	4,000,000	40%
KSE General Index	3.0	−2,000,000	−20%
MSM30 Index	4.0	1,000,000	10%
DSM20 Index	3.0	−1,000,000	−10%
SE All Share Index	2.0	3,000,000	30%
Correlation Factor	**LVaR (Crisis Market)**	**LVaR Volume**	**Expected Return**
ρ = Empirical	1,516,998	15.2%	0.069%
ρ = 1	2,171,834	21.7%	**Systematic Risk Factor (Beta)**
ρ = 0	1,280,591	12.8%	0.520

This table illustrates the asset allocation and composition of investable portfolio [3] using long- and short-sales trading positions under event (crisis) market conditions. The table also depicts LVaR under event market settings with different correlation factors, along with expected return and systematic risk factor (beta coefficient).

Source: Designed by the author using in-house developed software.

Table 1.5 Asset allocation and composition of investable market portfolio [4]

Market Index	Liquidation Period (in Days)	Market Value in AED	Asset Allocation per Market
DFM General Index	2.0	−2,000,000	−20%
ADSMI Index	3.0	3,000,000	30%
BA All Share Index	4.0	5,000,000	50%
KSE General Index	3.0	4,500,000	45%
MSM30 Index	4.0	5,000,000	50%
DSM20 Index	3.0	−3,000,000	−30%
SE All Share Index	2.0	−2,500,000	−25%
Correlation Factor	**LVaR (Crisis Market)**	**LVaR/Volume**	**Expected Return**
ρ = Empirical	1,583,396	15.8%	0.094%
ρ = 1	891,376	8.9%	**Systematic Risk Factor (Beta)**
ρ = 0	1,783,932	17.8%	−0.192

This table illustrates the asset allocation and composition of investable portfolio [4] using long- and short-sales trading positions under event (crisis) market conditions. The table also depicts LVaR under event market settings with different correlation factors, along with expected return and systematic risk factor (beta coefficient).

Source: Designed by the author using in-house developed software.

Figure 1.2 provides comparison of the empirical LVaR efficient frontier – under event (adverse) market settings – with the obtained investable portfolios defined using a 97.5% confidence level. As indicated above, the portfolio optimization and selection process is performed by allowing short-sales for the different stock markets of the GCC zone.

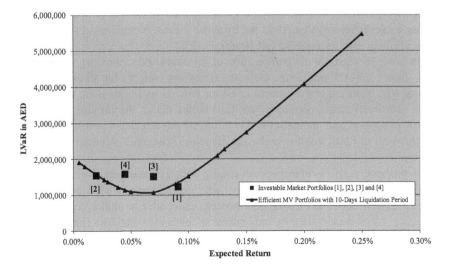

Figure 1.2 Efficient and investable market portfolios with LVaR technique

This figure depicts the empirical LVaR efficient frontier and investable portfolios under event market scenarios by relaxing short-sales constraint (i.e. allowing both long- and short-sales trading positions). As indicated, the four-benchmark investable portfolios [1], [2], [3] and [4] are located close to the efficient frontier, albeit to a different degree. For instance, investable portfolio [1] is located even lower than the efficient frontier, which can be regarded as a super-efficient portfolio. Furthermore, it seems that for long- and short-sales it is likely that the performance of efficient and investable portfolios are close to each other, however with different asset allocations.

Source: Designed by the author using in-house developed software.

In this particular case study, the weights are allowed to take negative or positive values; however, since arbitrarily high or low positive weights could have no financial investment sense, we determined to introduce lower and upper boundaries for the weights and in accordance with reasonable trading practices. Furthermore, for comparison purposes and since the endeavor in this work is to minimize LVaR under event market prospects subject to specific expected returns and total portfolio volume, we decided to plot LVaR versus expected returns and not the reverse, as it is commonly the norm in the various modern portfolio management literature.

In view of that, it is appealing to remark here that the four benchmark investable portfolios (that is, investable portfolios [1], [2], [3] and [4]) are positioned very close to the efficient frontier as indicated in Figure 1.2. It seems that by implementing the proposed optimization algorithm and portfolio selection procedure, it is possible to match more closely the risk-return characteristics of both efficient investable portfolios but with different asset allocations. In addition, it is important to state here that integrating the above realistic and restricted budget constraints into the portfolio manager's operational and selection process has caused investable portfolio [1] to have a tendency to be more efficient (or super-efficient) than other efficient portfolios dispensed on the efficient frontier. Therefore, it appears that by integrating the liquidity holding periods and other operational and financial constraints into the constrained optimization objective function, portfolio managers could use active trading strategies in order to earn excess returns that could be, by some means, beyond what the classical MV efficient frontier suggests (Al Janabi, 2012).

In order to illustrate the composition of investable portfolios [1], [2], [3] and [4], Tables 1.2–1.5 emphasize the asset allocation weights for all equity assets along with their expected returns, closeout periods and systematic risk factors (i.e. beta coefficient). Similarly, the four tables depict the minimum required LVaR to support and maintain the operations of these selected portfolios as well as the ratio of LVaR/volume under the assumption of three different correlation factors.

In this way, portfolio managers can employ appropriate downside-risk measures that allow them to take prompt decisions, which could produce a risk budget lower than a specific target. In this line of reasoning, under adverse market conditions for instance, LVaR is calculated by implementing downside volatilities (i.e. maximum negative assets returns throughout the sampling period). Thus, this analysis and the implemented LVaR algorithm permit one to determine the asymmetric aspect of risk and investable portfolios under adverse market conditions and, as such, it is a substantial improvement to the traditional Markowitz's MV approach.

Finally, Table 1.6 provides evidence of the asset allocations for the minimum-variance portfolio under Markowitz's MV assumptions and approach. It is evident from Table 1.6 that the MV optimizer can produce corner solutions and

Table 1.6 Asset allocation and composition of Markowitz's minimum-variance (MV) portfolio

Market Index	Liquidation Period (in Days)	Market Value in AED	Asset Allocation per Market
DFM General Index	10.0	−655,302	−7%
ADSMI Index	10.0	815,933	8%
BA All Share Index	10.0	6,845,667	68%
KSE General Index	10.0	1,884,033	19%
MSM30 Index	10.0	240,108	2%
DSM20 Index	10.0	472,988	5%
SE All Share Index	10.0	396,573	4%
Correlation Factor	**LVaR (Crisis Market)**	**LVaR/Volume**	**Expected Return**
ρ = Empirical	1,179,014	11.8%	0.070%
ρ = 1	1,884,796	18.8%	**Systematic Risk Factor (Beta)**
ρ = 0	1,056,473	10.6%	0.107

This table illustrates the asset allocation and composition of Markowitz's mean-variance (MV) portfolio using long- and short-sales trading positions under event (crisis) market conditions. We also use the regulatory parameterization of daily LVaR estimates as the benchmark in our MV analysis. The relevant LVaR parameterization liquidity risk factor is defined with 10 days (t = 10) holding period. Furthermore, the table depicts LVaR under event market settings with different correlation factors, along with expected return and systematic risk factor (beta coefficient).

Source: Designed by the author using in-house developed software.

meaningless asset allocation for portfolio selection. Indeed, it is well documented (Michaud, 1989) that MV optimizers, if left to their own devices, can sometimes lead to unintuitive portfolios with extreme positions in asset classes. In a portfolio optimization context, assets with large expected returns and low standard deviations will be overweighed and conversely, assets with low expected returns and high standard deviations will be underweighted (i.e. corner solutions and/or meaningless asset allocations).

Therefore, large estimation errors in expected returns and/or variances/covariances will introduce errors in the optimized portfolio weights (Fabozzi et al., 2006). As a result, these "optimized or efficient" portfolios are not necessarily well diversified and exposed to unnecessary ex post risk (Michaud, 1989). The reason for these phenomena is not a sign that MV optimization does not work but rather that the modern portfolio theory framework is very sensitive to small changes in the inputs. Simply put, the "optimized or efficient" portfolio, in many instances, is not "optimal" at all (Al Janabi, 2013; Michaud, 1989).

Concluding remarks and practical applications for portfolio selection and asset management

This review chapter fills an essential gap in the efficient portfolio selection literature by contributing a systematic modeling technique and optimization algorithms for trading risk with a dynamic asset allocation process and under the supposition of illiquid and event market settings, and within the environments of big data and expert systems. Moreover, in this chapter, the author develops a portfolio selection model and an optimization algorithm that allocates stock markets assets by minimizing the liquidity-adjusted value-at-risk (LVaR) objective function subject to applying credible operational and financial constraints, which are based on fundamental asset management considerations.

The proposed optimization algorithm demonstrates that better investable portfolios can be obtained than using the traditional Markowitz's mean-variance (1952) portfolio theory. The empirical results show that the obtained investable portfolios are located very close to the efficient frontier, but with more credible asset allocations than the traditional Markowitz's (1952) MV method. The empirical optimization results show that this alternate LVaR technique can be deemed as a robust portfolio selection and management tool and can have many uses and applications in real-world portfolio selection and asset management practices, mainly for asset managers with large financial (or other asset) portfolios.

The presented optimization algorithm for portfolio selection and empirical results of this review chapter can have important practical uses and applications for financial trading entities, financial institutions, risk managers, portfolio managers, regulators and policymakers operating in both emerging and developed markets, particularly in the wake of the most recent financial crisis.

For instance, risk managers and portfolio managers can use the proposed modeling technique and simulation algorithms for portfolio selection and management for the assessment of appropriate asset allocations of different structured investable portfolios. In fact, the proposed optimization algorithms have the potential of producing realistic risk-return profiles, may improve real-world understanding of embedded risks and asymmetric microstructure patterns, and could potentially create better investable portfolios and asset allocation for portfolio managers than the classical Markowitz's (1952) technique.

Finally, the optimization algorithms for portfolio selection and management and liquidity modeling framework can be beneficial also to asset management firms in emerging and developed economies, above all in emphasizing rational economic capital allocation in light of the repercussions of the latest subprime financial crunch.

Notes

1 In fact, the concept of liquidity risk in financial markets and institutions can imply either the added transaction costs related to trading large quantities of a certain financial security or with the ability to trade this financial asset without triggering significant changes in its market prices (see Roch and Soner (2013) for further details and empirical analysis).
2 For an excellent survey of recent contributions to robust portfolio strategies, from operations research and finance to the theory of portfolio selection, one can refer to Fabozzi et al. (2010). In addition, for important recent literature on loan portfolio optimization with genetic algorithms and for the optimization of bank lending decisions using genetic algorithm-based model, one can refer to Metawa et al. (2016) and Metawa et al. (2017).
3 For other relevant literature on liquidity, asset pricing and portfolio choice and diversification, one can refer to Angelidis and Benos (2006); Berkowitz (2000); Madhavan et al. (1997); Hisata and Yamai (2000); Le Saout (2002); Amihud et al. (2005); Takahashi and Alexander (2002); Cochrane (2005); Roll (1992); and Meucci (2009), among others. Furthermore, in their research paper, Sunny B. Walter Madoroba and Kruger (2015) introduce a new value at risk (VaR) model that incorporates intraday price movements on high–low spreads and adjusts for a trade impact measure, a novel sensitivity measure of price movements due to trading volumes. Furthermore, the authors compare and contrast 10 worldwide-recognized liquidity risk management models including the "Al Janabi Model," which is used in this chapter for liquidly risk modelling.
4 In his research papers, Al Janabi (2012 and 2014) has investigated and applied similar optimization algorithms for commodity and stock markets. While his general optimization elements and parameters are adapted herein, this chapter builds on Al Janabi's (2012, 2014) research papers and differs in the sense that it includes a detailed optimization algorithm approach for portfolio selection and research design that can be accessible to market practitioners, particularly commodity and stock market asset management entities. This chapter also includes in-depth analysis, discussion and justification of LVaR as a portfolio selection technique as well as further considerations and recommendations for asset management practical applications.
5 Indeed, the time-varying pattern of asset volatility has been widely recognized and modeled as a conditional variance within the GARCH framework as originally developed by Engle (1982, 1995). Engle (1982) introduced a likelihood ratio test to ARCH effects and a maximum likelihood method to estimate the parameters in the ARCH model. This approach was generalized by Bollerslev (1986) and Engle and Kroner (1995). In fact, the generalized autoregressive conditional heteroskedasticity in mean, the GARCH-M (1,1) model, is used in our empirical analysis for the estimation of expected return and conditional volatility for each of the time-series variables.

6 If the purpose of the risk analysis is to investigate diverse stock market dependences and related risk management measure, then $Asset_i$ should be the mark-to-market prices of the individual stock market indices.

7 In this chapter, the concept of investable market portfolios refers to rational financial portfolios that are subject to meaningful financial and operational constraints. In this sense, investable market portfolios lie off the efficient frontiers as defined by Markowitz (1952) and instead have logical and well-structured long- and short-asset allocation proportions.

8 The mathematical approach presented herein is largely drawn from Al Janabi's 2012, 2013 and 2014 research papers.

9 It is important to note that equation 1.11 can be used to calculate LVaR for any time horizon subject to the constraint that the overall LVaR figure should not exceed at any setting the nominal exposure, in other words the total trading volume.

10 In this class of models (that is, GARCH-M (1,1)), the conditional variance enters into the conditional mean equation as well as the usual error variance part. As such, when the return of a security is dependent on its volatility, one can use the GARCH-M (1,1) model formulation. Indeed, the GARCH-M (1,1) model implies that first, there exists serial correlation in the return series and second, these serial correlations are introduced into the volatility process due to a risk premium.

11 The historical database of daily indices levels is drawn from Reuters 3000 Xtra Hosted Terminal Platform and Thomson Reuters Datastream datasets.

12 In fact, there is no readily available time-series database on the liquidity of different stock markets. However, as indicated in equation 1.12 this can be estimated from the specific daily trading volume of any trading asset on a certain date and the historical average trading volume of that commodity for a certain period of time (for instance, the average daily/monthly trading volume for the last six months). In the real-world of financial markets, the number of days required to unwind a trading position can be assessed intuitively by individual traders based on their daily experience of how much volume of each trading asset they can liquidate (close out) into the market without disturbing market prices. For our purpose of introducing a robust portfolio selection and management methodology to a diverse portfolio of the GCC zone stock markets, the daily liquidation horizon for each stock market is included in parenthesis as follows:

(1) ADSMI Index [3.0]; (2) DFM General Index [2.0]; (3) KSE General Index [3.0]; (4) BA All Share Index [4.0]; (5) MSM30 Index [4.0]; (6) DSM20 Index [3.0]; (7) SE All Share Index [2.0]; (8) Shuaa GCC Index [1.0]; (9) Shuaa Arab Index [1.0].
       Furthermore, it is possible to run the optimization engine for portfolio selection in such a way to determine the optimal number of days for liquidation (threshold) by building a portfolio where one can specify the required asset allocations for each asset, expected return of the portfolio, risk parameters for each asset and any other operational and financial constraints the portfolio manager would like to enforce on the portfolio, and then solve for the optimal liquidation thresholds that can minimize the objective function with all its boundaries.

13 In fact, the Pearson correlation factors can be replaced by nonparametric dependence measures, such as, Kendal's tau and/or the different copulas functions.

14 The mathematical and optimization algorithms are largely drawn from Al Janabi (2012, 2013, and 2014) research papers.

15 Thus if trading volume is low because of a "one-way market," in that most people are seeking to sell rather than to buy, then t_i can rise substantially (Saunders and Cornett, 2008).

References

Abad, P., Benito, S. and Lopez, C. (2013), "A comprehensive review of value at risk methodologies," *The Spanish Review of Financial Economics*, Vol. 12, No. 1, pp. 15–32.

Acerbi, C. and Tasche, D. (2002), "On the coherence of expected shortfall," *Journal of Banking and Finance*, Vol. 26, No. 7, pp. 1487–1503.

Agarwal, V. and Naik, N. Y. (2004), "Risks and portfolio decisions involving hedge funds," *Review of Financial Studies*, Vol. 17, No. 1, pp. 63–98.

Al Janabi, M.A.M. (2012), "Optimal commodity asset allocation with a coherent market risk modeling," *Review of Financial Economics*, Vol. 21, No. 3, pp. 131–140.

Al Janabi, M.A.M. (2013), "Optimal and coherent economic-capital structures: Evidence from long and short-sales trading positions under illiquid market perspectives," *Annals of Operations Research*, Vol. 205, No. 1, pp. 109–139.

Al Janabi, Mazin A. M. (2014), "Optimal and investable portfolios: An empirical analysis with scenario optimization algorithms under crisis market prospects," *Economic Modelling*, Vol. 40, pp. 369–381.

Almgren, R. and Chriss, N. (1999), "Optimal execution of portfolio transaction," Working Paper, Department of Mathematics, The University of Chicago.

Amihud, Y., Mendelson, H. and Pedersen, L. H. (2005), "Liquidity and asset prices," *Foundations and Trends in Finance*, Vol. 1, No 4, pp. 269–364.

Angelidis, T. and Benos, A. (2006), "Liquidity adjusted value-at-risk based on the components of the bid-ask spread," *Applied Financial Economics*, Vol. 16, No. 11, pp. 835–851.

Artzner, P., Delbaen, F., Eber, J. M. and Heath, D. (1999), "Coherent measures of risk," *Mathematical Finance*, Vol. 9, pp. 203–228.

Bangia, A., Diebold, F., Schuermann, T. and Stroughair, J. (2002), "Modeling liquidity risk with implications for traditional market risk measurement and management." In Figlewski, S. and Levich, R. M. (Eds.), *Risk Management: The State of the Art, The New York University Salomon Center Series on Financial Markets and Institutions*, Vol. 8, pp. 3–13.

Berkowitz, J. (2000), "Incorporating liquidity risk into VaR models," Working Paper, Graduate School of Management, University of California, Irvine.

Best, M. J. and Grauer, R. R. (1991), "On the sensitivity of mean-variance-efficient portfolios to changes in asset means: Some analytical and computational results," *Review of Financial Studies*, Vol. 4, pp. 315–342.

Bollerslev, T. (1986), "Generalized autoregressive conditional heteroscedasticity," *Journal of Econometrics*, Vol. 31, No. 1, pp. 307–327.

Cochrane, J. H. (2005), *Asset Pricing*. Princeton, NJ: Princeton University Press.

Engle, R. F. (1982), "Autoregressive conditional heteroskedasticity," *Econometrica*, Vol. 50, No. 1, pp. 987–1008.

Engle, R. F. (1995), *ARCH Selected Readings, Advanced Texts in Econometrics*. Oxford: Oxford University Press.

Engle, R. F. and Kroner, K. (1995), "Multivariate simultaneous generalized ARCH," *Econometric Theory*, Vol. 11, No. 1, pp. 122–150.

Fabozzi, F. J., Focardi, S. and Kolm, P. (2006), "Incorporating trading strategies in the Black-Litterman framework," *Journal of Trading*, Spring Issue, pp. 28–37.

Fabozzi, F. J., Huang, D. and Zhou, G. (2010), "Robust portfolios: Contributions from operations research and finance," *Annals of Operations Research*, Vol. 176, No. 1, pp. 191–220.

Hisata, Y., and Yamai, Y. (2000), "Research toward the practical application of liquidity risk evaluation methods," Discussion Paper, Institute for Monetary and Economic Studies, Bank of Japan, p. 23.

Ho, L.-c., Cadle, J. and Theobald, M. (2008), "Portfolio selection in an expected shortfall framework during the recent 'credit crunch' period," *Journal of Asset Management*, Vol. 9, No. 2, pp. 121–137.

Gaivoronski, A. and Pflug, G. (2005), "Value-at-risk in portfolio optimization: Properties and computational approach," *Journal of Risk*, Vol. 7, No. 2, pp. 1–31.

Garcia, R., Renault, E. and Tsafack, G. (2007), "Proper conditioning for coherent VaR in portfolio management," *Management Science*, Vol. 53, No. 3, pp. 483–494.

Jarrow, R. and Subramanian, A. (1997), "Mopping up liquidity," *Risk*, Vol. 10, No. 12, pp. 170–173.

Jobson, J.D. and Korkie, B.M. (1981), "Putting Markowitz theory to work," *Journal of Portfolio Management*, Vol. 7, pp. 70–74.

Jorion, P. (1991), "Bayesian and CAPM estimators of the means: Implications for portfolio selection," *Journal of Banking and Finance*, Vol. 15, pp. 717–727.

Kaplanski, G. and Kroll, Y. (2002), "VaR risk measures versus traditional risk measures: An analysis and survey," *Journal of Risk*, Vol. 4, No. 3, pp. 1–27.

Le Saout, E. (2002), "Incorporating liquidity risk in VaR models," Working Paper, Paris 1 University.

Madhavan, A., Richardson, M. and Roomans, M. (1997), "Why do security prices change? A transaction-level analysis of NYSE stocks," *Review of Financial Studies*, Vol. 10, No. 4, pp. 1035–1064.

Madoroba, S. B. W. and Kruger, J. W. (2015), "Liquidity effects on value-at-risk limits: Construction of a new VaR model," *Journal of Risk Model Validation*, Vol. 8, pp. 19–46.

Mandelbrot, B. (1963), "The variation of certain speculative prices," *Journal of Business*, Vol. 36, pp. 394–419.

Markowitz, H. (1952), "Portfolio selection," *Journal of Finance*, Vol. 7, No. 1, pp. 77–91.

Metawa, N., Elhoseny, M., Hassan, K. and Hassanien, A. (2016), "Loan portfolio optimization using Genetic Algorithm: A case of credit constraints," Proceedings of 12th International Computer Engineering Conference (ICENCO), IEEE, 59–64.

Metawa, N., Hassana, M. and Elhoseny, M. (2017), "Genetic algorithm based model for optimizing bank lending decisions," *Expert Systems with Applications*, Vol. 80, pp. 75–82.

Meucci, A. (2009), "Managing Diversification," *Risk*, Vol. 22, No. 5, pp. 74–79.

Michaud, R.O. (1989), "The Markowitz optimization enigma: Is 'Optimized' optimal?" *Financial Analysts Journal*, Vol. 45, No. 1, pp. 31–42.

Morgan Guaranty Trust Company (1994), *Risk Metrics-Technical Document*. New York: Morgan Guaranty Trust Company, Global Research.

Poon, S.-H. and Granger, C. (2005), "Practical issues in forecasting volatility," *Financial Analysts Journal*, Vol. 61, No. 1, pp. 45–56.

Roch, A. and Soner, H.M. (2013), "Resilient price impact of trading and the cost of illiquidity," *International Journal of Theoretical and Applied Finance*, Vol. 16, No. 6, pp. 1–27.

Rockafellar, R.T. and Uryasev, S. (2000), "Optimization of conditional value-at-risk," *Journal of Risk*, Vol. 2, No. 3, pp. 21–41.

Rockafellar, R.T. and Uryasev, S. (2002), "Conditional value-at-risk for general loss distributions," *Journal of Banking & Finance*, Vol. 26, No. 7, pp. 1443–1471.

Roll, R. (1992), "A mean/variance analysis of tracking error," *Journal of Portfolio Management*, Vol. 18, pp. 13–22.

Ruozi, R. and Ferrari, P. (2013), "Liquidity risk management in banks: Economic and regulatory issues," *Springer Briefs in Finance*, http://doi.org/10.1007/978-3-642-29581-2.

Saunders, A. and Cornett, M. (2008), *Financial Institutions Management*, 6th Edition. Singapore: McGraw Hill International Edition.

Takahashi, D. and Alexander, S. (2002), "Illiquid alternative asset fund modeling," *Journal of Portfolio Management*, Vol. 28, No. 2, pp. 90–100.

Appendix A

Optimization algorithm for portfolio selection and research design for computer programming purposes

In this chapter, we discuss a model for optimizing portfolio risk-return with LVaR constraints using realistic operational and financial scenarios. In this appendix, we examine all the necessary stages for computer programming ends such that other researchers can have a useful tool to replicate the research design and optimization algorithms for portfolio selection and management.

Essentially, our proposed approach is a robust enhancement of the classic Markowitz (1952) MV approach, where the original risk measure (variance) is replaced by LVaR algorithms. The task is fulfilled by minimizing LVaR while requiring minimum expected returns, subject to putting into effect a number of financial and operational meaningful constraints under event market conditions. As a result, by allowing different expected returns we can select investable portfolios under adverse market outlooks. On the other hand, we can also maximize expected returns while not allowing for large exposure of risks.

Set against this background and to maximize its utility as a portfolio selection and management method, we have constructed the portfolio management tool such that the proposed risk-engine and robust scenario optimization algorithms proceed according to the following stages.

Stage 1 of the optimization algorithm for portfolio selection

In this first stage, we make an attempt to define a nonlinear dynamic risk-function (i.e. objective function) of multivariability. For this purpose, the nonlinear dynamic risk-function can be defined as a vector of (1) monetary investment in each asset class; (2) closeout periods (or unwinding periods) of each asset class and the overall closeout periods of investable portfolios; (3) overall trading volume of investable portfolios; (4) constrained asset allocation proportions according to contemporary financial market regulations and subject to the imposition of rational and meaningful operational and financial boundaries; (5) downside risk constraints so that additional risk resulting from any non-normality and illiquid assets may be used to estimate the characteristics of investable portfolio. This enables a much more generalized framework to be developed, with the distributional assumption most appropriate to the type of financial assets to be employed, and which can be of crucial value for more accurate market risk assessment during market stress periods and particularly when liquidity dries up; (6) correlation coefficients among all asset classes; (7) expected returns of investable portfolios; (8) confidence level of estimated parameters under different scenarios and market settings; and (9) portfolio managers' choices of a combination of long- and short-sales trading asset positions.

In this backdrop, the objective risk-function is defined and explained along these lines:

Portfolio trading risk in the presence of multiple risk factors is determined by the combined effect of individual risks. The magnitude of total risk is determined not only by the magnitudes of individual risk factors but also by their correlations and dependence measure function. In fact, portfolio effects are crucial in risk management not only for large diversified portfolios but also for individual instruments that depends on several risk factors (Al Janabi, 2013). For multiple assets, the LVaR of a portfolio of financial assets is a function of the individual risk of each asset, holding period (closeout or liquidation horizon) and the Pearson correlation (dependence) factors [$\rho_{i,j}$] between the returns on the individual securities, detailed as follow[13,14]

$$LVaR_{P_{adj}} = \sqrt{\sum_{i=1}^{n}\sum_{j=1}^{n} LVaR_{i_{adj}}\, LVaR_{j_{adj}}\, \rho_{i,j}} = \sqrt{\left[LVaR_{adj}\right]^{T}\left[\rho\right]\left[LVaR_{adj}\right]}$$

(1A.1)

In fact, equation 1A.1 is a generalized formula for the calculation of $LVaR_p$ for any portfolio regardless of the number of securities. It should be noted that the second term of this formula is presented in terms of matrix algebra techniques – a useful form to avoid mathematical complexity as more and more securities are added. This approach can simplify the algorithmic programming process and can permit a straightforward incorporation of long-only and long- and short-sales positions in the market risk assessment process.

Indeed, equation 1A.1 considers adverse price impact liquidity risk throughout the closeout period. As a result, the assumption of a given closeout horizon for orderly liquidation inevitably implies that asset liquidation occurs during the holding period. Accordingly, scaling the holding period to account for orderly liquidation can be justified if one allows the assets to be liquidated throughout the closeout period. In order to perform the calculation of LVaR under accurate illiquid market circumstances, we can define the following throughout the closeout period:

$$LVaR_{i_{adj}} = VaR_i\sqrt{\frac{(2t_i + 1)\,(t_i + 1)}{6t_i}}$$

(1A.2)

where VaR_i = value at risk of asset i under liquid market conditions; $LVaR_{i_{adj}}$ = value at risk of asset i under illiquid market conditions; t_i = number of trading days required for orderly liquidation of asset i. Moreover, the latter equation indicates that $LVaR_{i\,adj} > VaR_i$, however when the required number of days to liquidate the entire asset is one trading day, then we have a special case in which $LVaR_{i\,adj} = VaR_i$. Consequently, the difference between $LVaR_{i\,adj}$ and VaR_i should equal to the residual market risk due to the illiquidity of asset$_i$ under adverse market conditions. Furthermore, it is important to note that equation 1A.2 can be used for the calculation of LVaR for any time horizon subject to applying a constraint on the total LVaR figure. In fact, the overall LVaR figure should not exceed the nominal exposure at any setting, or in other words the total trading volume of the portfolio.

As a matter of fact, the number of liquidation days or closeout time (t_i) necessary to liquidate the entire position of asset i fully is related to the choice of the liquidity threshold; however, the size of this threshold is likely to change under severe market conditions. Effectively, a linear liquidation procedure of assets is assumed in equation 1.A.2, that is, selling equal parts of each asset every day until the last trading day (t_i), where the entire asset is sold. The above model is more appropriate for daily trading circumstances where traders can unwind part of their positions on a daily basis.

Indeed, the choice of liquidation horizon can be estimated from the total trading position size and daily trading volume that can be unwound into the market without significantly disrupting market prices; and in actual practices it is generally estimated as:

$$t_i = Total\ Trading\ Position\ Size\ of\ Asset_i\ /\ Daily\ Trading\ Volume\ of\ Asset_i$$
(1A.3)

As such, the closeout time (t_i) is the time required to bring the positions to a state where the financial entity can make no further loss from the trading positions. It is either the time taken to sell the long positions or alternatively the time required to buy securities in case of short positions. In real practices, the daily trading volume of any trading asset is estimated as the average volume over some period of time, generally a month of trading activities. In effect, the daily trading volume of assets can be regarded as the average daily volume or the volume that can be unwound under severe critical situation periods.[15] The trading volume in a crisis period can be roughly approximated as the average daily trading volume less a number of standard deviations. Although this alternative approach is quite simple, it is still relatively objective. Moreover, it is reasonably easy to gather the required data to perform the necessary liquidation scenarios.

Against this backdrop we can now define VaR_i for single asset positions. By and large, the absolute value of VaR_i for any single financial asset can be defined in monetary terms as:

$$VaR_i = \left| (\mu_i - \alpha * \sigma_i)[Asset_i * Fx_i] \right| \approx \left| \alpha * \sigma_i[Asset_i * Fx_i] \right|$$
(1A.4)

where μ_i is the expected return of asset$_i$, α is the confidence level (or in other words, the standard normal variant at confidence level α) and σ_i is the forecasted standard deviation (or conditional volatility) of the returns of asset$_i$. While the term $Asset_i$ denotes the mark-to-market monetary value, Fx_i indicates the unit foreign exchange rate of $Asset_i$. Without a loss of generality, we can assume that the expected value of daily returns μ_i is close to zero. As such, although equation 1A.4 includes some simplifying assumptions, it is routinely used by researchers and practitioners in financial markets for the estimation of VaR for any single trading position.

Stage 2 of the optimization algorithm for portfolio selection

In the second stage, we define the corresponding robust scenario optimization algorithm, which is based on quadratic programming techniques, subject to applying meaningful financial and operational meaningful constraints. Furthermore, in the development of portfolio investment policy there are many types of constraints that can be considered as an integral part of the optimization process. These constraints are drawn from rational financial investment considerations and can be used in various applications to bound percentiles of loss distributions, and it can include constraints on the asset classes that are allowable and concentration limits on investments. Moreover, in making the asset allocation decision, consideration must be given to any risk-based capital requirements. For this objective, the scenario optimization algorithms can be defined as a minimization algorithmic process of the dynamic objective risk-function and as follows:

(1) The minimization process is attained here by minimizing the objective risk-function while requiring minimum expected returns subject to imposing several rational and meaningful financial and operational constraints; (2) at a minimum, the bounding limits may include the following: (a) the target investable portfolio's expected return; (b) total volume of the investable portfolio; (c) monetary asset allocation of each asset class; (d) portfolio managers' choices of long-only positions or a combination of long- and short-sales trading positions; and (e) closeout or unwinding liquidity horizons of each asset class.

In essence, our suggested optimization algorithm and portfolio selection method is a robust improvement to the classical Markowitz mean-variance approach, where the original risk measure, variance, is replaced by the LVaR algorithm. The task is attained here by minimizing the LVaR algorithm while requiring a minimum expected return subject to applying meaningful financial and operational constraints. Thus by considering different expected returns, we can generate investable portfolios using LVaR algorithm. Alternatively, we can also maximize expected returns while not allowing for large risks. For the purpose of this research study, the optimization problem is formulated as follows:

From equation 1A.2, we can define liquidation horizon factor (LHF_i) for each trading asset as:

$$LHF_i = \sqrt{\frac{(2t_i + 1)(t_i + 1)}{6t_i}} \geq 1.0 \; ; \forall_i \quad i = 1, 2, ..., n \qquad (1A.5)$$

After substituting equation 1A.5 into equation 1A.1, we can now compute the minimum portfolio downside risk by solving the following nonlinear quadratic algorithm (objective function) under adverse and illiquid market situations:

$$Minimize : LVaR_{P_{adj}} = \sqrt{\sum_{i=1}^{n} \sum_{j=1}^{n} LVaR_{i_{adj}} LVaR_{j_{adj}} \rho_{i,j}}$$

$$= \sqrt{\left[LVaR_{adj} \right]^{T} \left[\rho \right] \left[LVaR_{adj} \right]} \qquad (1A.6)$$

The above objective function can be minimized conditional on satisfying the following operational and financial budget constraints as specified by the portfolio manager:

$$\sum_{i=1}^{n} R_i x_i = R_P \; ; l_i \leq x_i \leq u_i \quad i = 1,2,....,n \qquad (1A.7)$$

$$\sum_{i=1}^{n} x_i = 1.0 \; ; l_i \leq x_i \leq u_i \quad i = 1,2,...., n \qquad (1A.8)$$

$$\sum_{i=1}^{n} V_i = V_P \quad i = 1,2,...., n \qquad (1A.9)$$

$$\left[LHF \right] \geq 1.0 \; ; \forall_i \quad i = 1,2,....., n \qquad (1A.10)$$

Here R_P and V_P denote the target portfolio mean expected return and total portfolio volume, respectively, and x_i the weight or percentage asset allocation for each asset. The values l_i and u_i, $i = 1, 2, . . ., n$, denote the lower and upper constraints for the portfolio weights x_i. If we choose $l_i = 0$, $i = 1, 2, . . ., n$, then we have the situation where no short-sales are allowed. Moreover, $\left[LHF \right]$ indicates an $(n - 1)$ liquidity risk factor vector for all $i = 1, 2, . . ., n$.

Now the portfolio manager can specify different liquidity horizons and correlation factors and calculate the necessary amount of LVaR to sustain trading operations without subjecting the financial entity to insolvency matters. In fact, the rationale behind imposing the above constraints is to comply with contemporary regulations that enforce capital cushion requirements on investment companies, proportional to their LVaR and economic capital, besides other operational limits (volume trading limits, the choices of long/short-sales positions, etc.).

Stage 3 of the optimization algorithm for portfolio selection

In this final phase, we validate and compare the output results of investable portfolios obtained in Stage 2 with the requirements of investable portfolios defined in Stage 1. To this end, we attempt to rerun the optimization engine until a new convergence to meaningful investable portfolios is attained. At this stage, new investable portfolios with meaningful asset allocations structures, which satisfy the boundary conditions defined in Stage 2, are achieved accordingly.

2 Financial analysis for mobile and cloud applications

Jennifer Brodmann and Makeen Huda

Introduction

Cloud computing is one of the fastest-growing segments in the information technology (IT) industry (Subashini and Kavitha, 2011). This new technology is now being adopted in the finance industry due to its potential to increase capability and capacity without needing to invest in new technology infrastructure, personnel training, or software licensing. Cloud computing enables firms to share information with other departments and streamline processes to save time, money, and resources. Mobile applications are widely used in cloud environments to allow users to access information both on-site and remotely through mobile devices and computers.

The formal definition of cloud computing, according to the National Institute of Standards and Technology (NIST), is "a model for enabling ubiquitous, convenient, on-demand network access to a shared pool of configurable computing resources (e.g., networks, servers, storage, applications and services) that can be rapidly provisioned and released with minimal management effort or service provider interaction."[4] NIST states that there are five essential characteristics of cloud computing: "on-demand self-service, broad network access, resource pooling, rapid elasticity or expansion, and measured service." NIST also lists three "service models" (software, platform, and infrastructure) and four "deployment models" (private, community, public, and hybrid) that together categorize ways to deliver cloud services.

The modern world has been shaped by steady technological advances that have resulted in information occupying an increasingly essential place in society. These technological breakthroughs allow for the computation of immensely large sets of data, which makes cloud computing an important part of the economy.

While the rise of big data can be felt throughout society, the main big data drivers are science, the internet, finance, mobile devices, sensors, radio-frequency identification (RFID), and streaming data (Elhoseny, Abdelaziz et al., 2018). There is a growing trend of firms moving their finance and accounting software to cloud-based applications, with practicality in mind (Gill, 2011; KPMG, 2012). Cloud computing has the potential to build business systems and bring lower costs and higher profits (Zhang et al., 2010). Gill (2011) elaborates on this, stating "[n]ow the cloud application wave has reached the finance organization where it promises the same impact – lower cost, easier collaboration, and faster innovation." Smith (2011) states that finance is the most recent industry to adopt cloud computing, with the New York Stock Exchange (NYSE) announcing that they have adopted a financial services cloud. The financial services and banking industry has been using cloud computing, including such business process networks as SWIFT, which supports a community cloud approach. Gill (2011) discusses a recent survey that NetSuite conducted with approximately 800 Institute of Management Accountants (IMA) members, which asked finance professionals "What do you perceive as the single key benefit of moving your financials to the cloud?" The respondents offered six perceived benefits of using cloud computing, with the majority touting cloud computing's potential to increase accessibility of data and decrease total ownership costs.

Gupta et al. (2012) discuss cloud-based data management applications to analyze big data and suggest ways such applications can be used in business. They propose an example of how cloud computing can be used in finance for credit card fraud detection. They point out that the only real way to identify credit card fraud is to analyze consumers' spending patterns. These patterns can encompass a broad range of information, such as items purchased, geographical location, time frame of purchase, and so forth. As the number of credit card transactions is very high, keeping track of all this data is a daunting task that typical warehouse-based data processing is not well suited to do. Thus, using cloud computing is the best way of implementing the data mining algorithms that can most quickly and effectively catch credit card fraud.

Brandas et al. (2015) examine how cloud applications can be used in accounting and finance, specifically examining cloud computing's effect on accounting information systems (AIS). They note that cloud computing and mobile technologies are being increasingly adopted by small and medium-sized enterprises (SMEs). Large companies are also regularly making use of private cloud and hybrid cloud computing. The authors conducted a SWOT (strengths, weaknesses, opportunities, threats) analysis on these applications and concluded that cloud adoption provides scalability, mobility, and a reduction in maintenance costs, but the emphasis should be placed on data and application security since the application is storing sensitive and confidential data.

Ahmed et al. (2012) state that cloud computing comprises the application, storage, and connectivity. Each of these three segments has a different purpose and provides different products for businesses. They also extol cloud computing's potential to improve the energy efficiency of businesses' IT operations. In fact, energy costs account for over half of the operational expenses of running data centers. Cloud computing can offer novel methods, such as server consolidation and construction of networks of smaller and decentralized data centers, to reduce operational costs while maintaining performance.

Carcary et al. (2013) conduct a survey on cloud adoption by SMEs in Ireland and find that just under half of the respondents had adopted cloud computing in their business. Of these cloud computing adopters, 18% reported that finance was the key business function being migrated to the cloud. Further, 27% of the non-adopters reported that the main reason for their reluctance to implement cloud computing was insufficient financial resources, suggesting that a more in-depth assessment of cloud computing that acknowledges its potential financial benefits could sway non-adopters to become adopters. Khajeh-Hosseini et al. (2010) propose that third-party cloud infrastructure can improve income management and outflows for both finance staff and customers. Migrating a business's IT operations to a third-party cloud infrastructure as opposed to using an in-house IT department results in much smaller fixed and monthly costs as well as lower energy expenditure. Lagar-Cavilla et al. (2009) evaluate the adoption of SnowFlock, a cloud-based application that uses virtual machine technology, in quantitative finance using Quantlib, an open source toolkit for modeling stock trading, equity option pricing, and risk analysis. They find that virtualization in cloud computing can greatly increase performance, parallel computing, and scalability, all while reducing resource consumption.

In order to determine the various types of cloud applications that can be adopted by businesses, Chang et al. (2013) discuss the Cloud Computing Business Framework (CCBF), which provides a roadmap for organizations to adopt cloud computing. They find that the CCBF can provide benefits to organizations in the following dimensions: classification, organizational sustainability modeling, service portability, and linkage. These benefits can manifest in many ways, such as cost reductions, added value, enhanced business performance, better integration across business units, improved tracking of project status, and so forth.

Chang et al. (2012) study financial enterprise portability, which is the movement of application services from desktops to cloud applications as well as between clouds, where users can access the applications as if on the systems to which they are accustomed. Their study tests how financial applications can be used in the cloud. Their results indicate the cloud access use of workflow reflects the risks associated with business processes so that risk tolerance can be developed and analysis can lead to better decision-making. Implementation of File System as a Service (FSaas) can allow for faster calculations of pricing as well as more accurate risk modeling in order to meet the demands of academic and industry research related to finance. They also state that enterprise portability can reduce time and cost and improve performance.

Gai (2014) studies cloud computing in financial service institutions and states that adoption of cloud computing can improve four main domains for financial institutions: services, finance, marketing, and operations. He finds that cloud computing is more accepted by smaller financial service institutions than larger ones, although there is a general trend in the financial services industry toward increased use of cloud applications. Further, he finds that financial institutions can use private cloud-based services to reduce cost, increase asset utilization, and attract more attention from customers.

Types of cloud and mobile applications

Cloud and mobile applications encompass a wide array of categories and types. This section discusses the different types of cloud applications that are used in the finance industry. The types of cloud applications are referenced from Garg (2011) and Mell and Grance (2011).

Software as a service (SaaS)

SaaS are software products provided through a platform. Enterprise resource planning (ERP) software provides to firms a main database that supports multiple departments (e.g. accounting and finance), where these two departments can use the same information for the different needs of the firm. ERP software can also provide synchronized reporting and automation. It can streamline processes by bringing sales orders into the financial system to close out sales without having to rekey information. This results in more efficient order management. In addition, the unified database allows employees to track business performance. Consumers do not administer the cloud infrastructure, which includes the cloud network, cloud servers, cloud operating systems, data storage, and the capabilities of each application, except for specific settings that are user-specific. Cloud computing, also known as software as a service (SaaS) is promoting the growth of ERP software and is being used more by high-growth and mid-sized businesses.

Platform as a service (PaaS)

PaaS are platform clouds that help developers create as well as host web applications through other applications or data centers. Examples are Microsoft Azure and GoogleApp Engine. The consumer does not administer the underlying cloud infrastructure, which includes the cloud network, cloud servers, cloud operating systems, and data storage. However, customers can control the applications that are deployed as well as any user-specific settings for the environment that hosts the application.

Infrastructure as a service (IaaS)

IaaS are infrastructure clouds that provide hosting, storage, and networking capacity. This allows consumers to both deploy as well as run their arbitrary software, such as operating systems and applications. The consumer does not administer or control the underlying cloud infrastructure yet has control over the cloud operating systems, data storage, and any applications that are deployed. Consumers may also have some control over certain networking components, such as firewalls. NYSE Euronext CMCP is a type of IaaS and provides services for NYSE Euronext customers that use their financial services.

Business process as a service (BPaaS)

BPaaS are online business processes that help with administrative and expense functionality. Examples of BPaaA are ADP Employease, which offers "online business process services for HR, benefits administration and outsourcing," and AMEX Concur, which is "an online business process that connects travel suppliers and mobile solutions from around the world to provide advanced travel and expense functionality" (Garg, 2011).

Data as a service (DaaS)

DaaS are cloud-based data platforms and services. Examples of DaaS include Google Public Data, which creates large datasets that are simple to view, use, and communicate with; and Xignite Capital Markets Data, which is a platform that provides financial market data on demand.

Deployment clouds

Public cloud

Public clouds are accessed over the internet and are accessible to everyone who has an internet connection. Cloud providers manage and own all aspects of the application from the operations, facilities, and computing resources. Examples of public clouds are Amazon EC2 and Microsoft Azure.

Private cloud

Private clouds are accessible to only trusted users from a specific organization. This can all be managed by either the organization or the cloud provider. A private cloud may exist either on or off work organization premises.

Community cloud

Community clouds are, true to their name, all accessible to the members of a community, which can consist of certain organizations or groups. Co-management is done by partner organizations as well as the cloud provider. These may be owned and operated by the community or group organization, a third party, or a combination thereof.

Hybrid cloud

Hybrid clouds are a mixture of public, private, or community clouds that address the issue of solely public, private, or community cloud applications. These clouds are bound by technology that allows for the portability of both data and applications.

Main characteristics of cloud computing

On-demand self-service

Cloud computing allows for unilateral provisioning of computing resources, which include server time, storage, or network bandwidth. This does not require any human interaction with service providers.

Universal network access

This is also known as broad network access. The cloud provides system access for any user location or device through a laptop, tablet, mobile phone, or any device that has internet access.

Resource pooling

Cloud computing allows for resource sharing to reduce costs across users, with various physical and visual resources assigned and reassigned for current user demand. Types of resources shared include data storage, data processing, memory, and network bandwidth.

Rapid elasticity

Quick scaling up or down of capabilities can be done through cloud computing in real time, depending on the demand. This capability can be conducted in unlimited amounts in any range of time.

Pay per use

This is also known as measured service. Pay per use allows for the user to pay for only the resources they use, such as data storage, data processing, and user bandwidth.

Main issues of cloud and mobile applications in finance

In the survey used in Gill (2011), IMA members were asked about their concerns with cloud computing. We expand upon his findings and discuss these types of concerns in the following sections.

Customization

Respondents stated that customization, or the lack thereof, can be a concern for them in adopting cloud-based platforms. This may stem from businesses needing specific functionality for their operations that cannot be achieved without further investment in customized cloud-based applications, which may be cost prohibitive for small businesses. As Carcary et al. (2013) state, firms are reluctant to adopt cloud computing due to the financial costs. These costs may also be associated with the need for additional customization to suit a firm's needs.

Reliability

Availability of cloud-based applications is also a topic of concern. If an outage occurs in the cloud, this will result in users not having access to their applications. Typically, cloud vendors offer a 99.5% or more service level commitment to their users. In addition, through economies of scale cloud vendors can invest in several levels of failover and redundant systems to reduce system outages. Cloud computing users can also be subject to a "noisy-neighbor" effect, when a neighbor's applications can use an outsized share of computing resources, such as processing and memory, which then takes away these resources from the user. Cloud security providers are not required to share this information.

Application ownership

Application ownership is also a major concern because users may spend extensive resources in customizing the cloud application to suit their needs only to ultimately not have any ownership rights to the application. Specific sections of the contract between the user and the cloud service provider should be devised to address this issue.

Data ownership

Gray (2013) discusses the legal ownership of data in cloud computing with IT attorney Marcus Lee from the law firm Moore & Van Allen. Lee states that provisions regarding data ownership need to be specifically detailed within the cloud computing contract between the service provider and user. In particular, conditions regarding data encryption, security procedures, government subpoena of cloud data, and destruction of data at the termination of the contract should be spelled out in detail.

Security

Financial services industries are interested in adopting this cloud and mobile, but they are wary of subjecting their customers' information to security risk (Gill et al., 2011). Several academic studies have examined the security risks that face cloud computing (Sumter, 2010; Shaikh and Haider, 2011; Munir and Palaniappan, 2012). Subashini and Kavitha (2011) focus on studying the cloud computing security issues in the context of service delivery models. Despite having a substantial capacity to store data, the cloud environment can bring security concerns from users not knowing the exact location of the data (Odun-Ayo et al., 2017).

Munir and Palaniappan (2012) state that security and privacy problems are the most substantial challenge that cloud computing faces and this stems from its "multi-tenancy nature and the outsourcing of infrastructure, sensitive data and critical applications." They recommend using the "defense in depth security approach" to secure applications and data in cloud environments. This approach utilizes firewalls, email gateways, virtual private networks, virtual local area network segmentation, authentication, and intrusion detection systems to improve data integrity while protecting against security breaches and malware.

Shaikh and Haider (2011) maintain that the security concerns inherent in cloud computing are the only obstacles standing in the way of its widespread implementation. They survey the recent literature on the security threats that users face when adopting cloud computing and identify the most important ones to be "data loss, leakage of data, client's trust, user's authentication, malicious users handling, wrong usage of cloud computing and its services, and hijacking of sessions while accessing data." Data security is a major concern because of data breaches, especially since firms can be exposed to legal repercussions from a data breach or technical problem. To remedy these concerns, Shaikh and Haider recommend the use of a free, publicly available downloadable set of tools issued by the non-profit Cloud Security Alliance, designed "to help organizations create public and private clouds that comply with industry standards for accepted governance, risk, and compliance best practices."

Concerns with data security are based on cloud data centers and cloud applications being accessible over the internet, yet the most sensitive transactions are already being facilitated over the internet, such as online banking and payroll (Gill, 2011). In actuality, internet security with cloud applications use banking-level 128-bit SSL security, which is more heavily encrypted compared to a traditional local area network (LAN) used prior to the internet. Also, a firm having its financials hosted on a data center is a topic of concern, yet firms and their business applications are typically connected to the internet frequently, which can leave a firm vulnerable to hacking.

Ko et al. (2011) point out that the finance industry already has regulations, such as the Payment Card Industry Data Security Standard, that restrict the use of public clouds due to data protection issues. They propose a HyBrex (Hybrid Execution) model, which utilizes a public cloud for non-sensitive data and a private cloud for sensitive data, and thus partitions data for privacy, confidentiality, and security.

Conclusion

Cloud applications are becoming widely used in the finance industry for a variety of reasons. They provide a means for firms in the finance industry to streamline processes, efficiently share information with other departments, and utilize financial models in a fast and proficient manner. Zhang et al. (2010) discuss the main future trends of cloud computing, which include an expansion of cloud computing with an emphasis on open platforms; increased investment by Google allowing more business users to utilize Google Apps; the first set of SaaS 1.0 firms confronting bankruptcy risk; firms increasingly discontinuing the use of their own servers; the increasing use of private cloud computing; SaaS moving into the arena of business intelligence (BI); SAP and Oracle entering into the market of platform as a service (PaaS); and the rapidly increasing utilization of social networks in cloud computing. Gill (2011) states that a firm may benefit from cloud computing in several ways, including "improved collaboration, easier global delivery, lower total cost of ownership (TCO), and always-up-to-date accounting software." However, there are also issues that need to be addressed when utilizing cloud-based applications. These issues mainly concern data and application security. Firms should be vigilant about maintaining the security and confidentiality of client and firm data.

Currently there is limited research on financial analysis in cloud computing. Studies have discussed the trends and benefits of cloud computing, but there has not yet been an in-depth study of how cloud computing affects financial analysis specifically. There are several directions for future research in this area of interest. These include the types of cloud applications that can be used for financial analysis, the benefits and risks associated with financial analysis and cloud applications, and comparative studies of firms that have adopted cloud applications and firms that have not. In addition, since security is a primary concern for firms adopting cloud computing, a study examining the financial implications of security breaches using cloud computing would also be a great direction for future research. We think that the rising trend of cloud computing adoption in the financial industry will beget more research in this area.

Note

1 Brown, E.A. (2018, January 08). Final Version of NIST Cloud Computing Definition Published. Retrieved May 7, 2018, from www.nist.gov/news-events/news/2011/10/final-version-nist-cloud-computing-definition-published.

References

Ahmed, M., Chowdhury, A.S.M.R., Ahmed, M., & Rafee, M.M.H. (2012). An advanced survey on cloud computing and state-of-the-art research issues. *IJCSI International Journal of Computer Science Issues*, *9*(1), 1694–1814.

Brandas, C., Megan, O., & Didraga, O. (2015). Global perspectives on accounting information systems: Mobile and cloud approach. *Procedia Economics and Finance*, *20*, 88–93.

Brown, E.A. (2018, January 08). Final Version of NIST Cloud Computing Definition Published. Retrieved May 7, 2018, from www.nist.gov/news-events/news/2011/10/final-version-nist-cloud-computing-definition-published.

Carcary, M., Doherty, E., & Conway, G. (2013). The adoption of cloud computing by Irish SMEs-an exploratory study. *Electronic Journal of Information Systems Evaluation*, *16*(4), 258.

Chang, V., Li, C.S., De Roure, D., Wills, G., Walters, R. J., & Chee, C. (2012). The financial clouds review. *Cloud Computing Advancements in Design, Implementation, and Technologies*, *125*.

Chang, V., Walters, R. J., & Wills, G. (2013). The development that leads to the Cloud Computing Business Framework. *International Journal of Information Management*, *33*(3), 524–538.

Elhoseny, M., Abdelaziz, A., Salama, A., Riad, A., Sangaiah, A., & Muhammad, K. (2018). A hybrid model of internet of things and cloud computing to manage big data in health services applications. *Future Generation Computer Systems*, *86*, 1383–1394 DOI: https://doi.org/10.1016/j.future.2018.03.005.

Gai, K. (2014). A review of leveraging private cloud computing in financial service institutions: Value propositions and current performances. *International Journal of Computer Applications*, *95*(3).

Garg, A. (2011). Cloud computing for the financial services industry. *Sapient Global Markets White Papers*, 1–16.

Gill, A. Q., Bunker, D., & Seltsikas, P. (2011, December). An empirical analysis of cloud, mobile, social and green computing: Financial services IT strategy and enterprise architecture. In *Dependable, Autonomous and Secure Computing (DASC), 2011 IEEE Ninth International Conference on* (pp. 697–704). IEEE.

Gill, R. (2011). Why cloud computing matters to finance. *Strategic Finance, 92*(7), 43–48.

Gray, P. (2013, March 5). Legal issues to consider with cloud computing. Retrieved May 20, 2018, from www.techrepublic.com/blog/tech-decision-maker/legal-issues-to-consider-with-cloud-computing/.

Gupta, R., Gupta, H., & Mohania, M. (2012, December). Cloud computing and big data analytics: what is new from databases perspective? In *International Conference on Big Data Analytics* (pp. 42–61). Berlin, Heidelberg: Springer.

Khajeh-Hosseini, A., Greenwood, D., & Sommerville, I. (2010, July). Cloud migration: A case study of migrating an enterprise it system to IAAS. In *Cloud Computing (CLOUD), 2010 IEEE 3rd International Conference on* (pp. 450–457). IEEE.

Ko, S. Y., Jeon, K., & Morales, R. (2011). The HyBrex model for confidentiality and privacy in cloud computing. *HotCloud, 11*, 8–8.

KPMG. (2012, December). Have you looked at Finance & Accounting in the Cloud? Perhaps you should. Retrieved April 10, 2019, from https://home.kpmg/content/dam/kpmg/pdf/2013/03/have-you-looked-at-finance-accounting-cloud.pdf

Lagar-Cavilla, H. A., Whitney, J. A., Scannell, A. M., Patchin, P., Rumble, S. M., De Lara, E., . . . & Satyanarayanan, M. (2009, April). SnowFlock: rapid virtual machine cloning for cloud computing. In *Proceedings of the 4th ACM European Conference on Computer Systems* (pp. 1–12). ACM.

Mell, P., & Grance, T. (2011). The NIST definition of cloud computing (NIST Special Publication 800–145). National Institute of Standards and Technology, Tech. Rep. Retrieved April 10, 2019, from https://nvlpubs.nist.gov/nistpubs/Legacy/SP/nistspecialpublication800-145.pdf

Munir, K., & Palaniappan, S. (2012). Security threats/attacks present in cloud environment. *IJCSNS, 12*(12), 107.

Odun-Ayo, I., Omoregbe, N., Odusami, M., & Ajayi, O. (2017, December). Cloud ownership and reliability – Issues and developments. In *International Conference on Security, Privacy and Anonymity in Computation, Communication and Storage* (pp. 231–240). Cham: Springer.

Shaikh, F. B., & Haider, S. (2011, December). Security threats in cloud computing. In *Internet Technology and Secured Transactions (ICITST), 2011 International Conference for* (pp. 214–219). IEEE, Abu Dhabi, United Arab Emirates.

Smith, D. M. (2011). *Hype Cycle for Cloud Computing, 2011.*Stamford: Gartner Inc., 71.

Subashini, S., & Kavitha, V. (2011). A survey on security issues in service delivery models of cloud computing. *Journal of Network and Computer Applications, 34*(1), 1–11.

Sumter, L. Q. (2010, April). Cloud computing: Security risk. In *Proceedings of the 48th Annual Southeast Regional Conference* (p. 112). ACM.

Zhang, S., Zhang, S., Chen, X., & Huo, X. (2010, January). Cloud computing research and development trend. In *Future Networks, 2010. ICFN'10. Second International Conference on* (pp. 93–97). IEEE, Washington DC, USA.

3 Eye-movement study of customers on video advertising marketing

Ruoqi Liang and Xiaolong Liu

Introduction

Video advertisements (ads) now appear before, during or after streaming game or animation contents as in-webpage video advertising. However, the effectiveness of video ads on consumer cognitive processes remains uncertain. Existing visual marketing literature focuses mainly on how salient factors such as ad location, format, size and presentation duration influence video marketing effectiveness [1], the relationship between attention to visual key elements (i.e. products, brands and endorser elements) and ad effectiveness has not been investigated. Therefore, this paper explores the relationships between these key ad elements and ad effectiveness for video ads to help the decision making in advertising business intelligence.

While attention is a gate through which information enters to reach cognitive processes of increased interest [2], tracking visual attention for elements can be regarded as obtaining the most direct information to bring out consumers' inner cognition to the video ad [3]. Traditional advertising research mainly uses self-report measurements to evaluate the video ad [4]. However, it is difficult to understand the effect of each advertising element on consumers' cognition and emotion using a self-report, because individuals typically assess the ad in its entirety. Using the eye-tracking method, this study enables researchers to learn the impact of each element on consumer persuasion. Moreover, due to advancements in eye-tracking technology, gaze tracking has become a very useful tool in the indirect measurement of attention and cognition [5–6]. The ability to track these movements can yield meaningful insights into unconscious cognitive activity, which can often be very difficult to tap with self-report attention. This study thus fills the existing research gaps by demonstrating the feasibility of objective measures of attention, using an eye-tracking method, to help explore the relationship between ad elements and video advertising effectiveness. Attention is limited and selected, which means a lot of attention is invalid to have an impact on consumer decision [7]. The message may or may not remain in the consumer's memory after cognitive processing. Thus, this study helps define meaningful eye movement with deeper cognitive level when exposing video ad.

Recent research focused on a set of core constructs that can affect advertising success, such as attention, memory, affect and desirability. In this study, three ads' effectiveness that have been shown to reliably tap into consumers' constructs are addressed: subjective reported attention, memory (ad recall), and affect (attitude toward ad) [8]. Advertising researchers refer to the attention aspect to attract focus to an ad. In this study, subjective attention is defined as the consumers' focus attention on the overall ad reported by themselves, while objective attention is defined as the eye-movement tracking to the video ad. Advertising research focused on the memory aspects to evaluate the quality of ads. Literature shows that people who recalled ads were more likely to have favorable outcomes than those who reported less exposure [9]. Attitude is considered to be one of the key determinants of advertising efficiency. Attitude toward advertising was considered as an overall feeling toward advertising of video in general.

We focus on tracking three highly relevant visual elements (products, branding and endorser elements), which are more likely to acquire visual attention [10] in video ads, to explore ad effectiveness. Moreover, demographic characteristics including gender and age [11] as well as variables including product involvement [12], brand familiarity and endorser familiarity [13] are used as control variables to explore video ad effectiveness because previous research has found that the effect of these variables is significant in relation to ad effectiveness. The researcher answers the following questions using objective eye-movement data to key elements and subjective ad effectiveness measures, analyzed using three multiple logistic regression models:

- Question 1: How do eye movements on product, brand and endorser relate to consumers' reported ad attention (RAT)?
- Question 2: How do eye movements on product, brand and endorser relate to consumers' recall for ad (AR)?
- Question 3: How do eye movements on product, brand and endorser relate to consumers' attitude toward ad (ATA)?

Research methodology

Materials

Ads from different regions were presented to participants, and we expected the participants to be unfamiliar with the ads. Six actual 30-second commercials were chosen for a range of products including shoes, liquor, mobile phones, clothes, humidifiers and razors. All ads have the product, brand and endorser elements. Different products were included to improve the validity of the results, and the length of the ads was consistent with the length of typical online video ads used on websites. To control for primacy and the recency effect, different presentation orders for the six video ads were created for each participant. Using the random number function in Excel, each ad was placed into an ad position (A1–F6) for each rotation.

Participants

Of 73 paid participants in the controlled experiment, ten participants were excluded because their eye-tracking calibrations failed to meet the required accuracy and two were excluded due to a lack of questionnaire data. The final sample used in the analysis comprised 61 students and staff members (28 men; mean age = 25 years, range = 17–46 years). Each participant browsed six video ads and completed the eye-tracking studies and the questionnaire. While the sample size is small, it is adequate for this study because six repeated measures were gathered from each participant, allowing the power to detect large-sized effects (Cohen's $f = 0.4$) at $p < .05$ with 100% power [14]. Participants were in the target audience for the video ads.

Process

The participants were required to answer the product involvement measures for the six product categories in advance, corresponding to the ads used in this study. Our lab assistant briefed the participants and obtained their signed informed consent before sitting the participants in front of a computer connected to an Eye Tribe Tracker [15]. During the experiment, the participants were exposed to six video ads while their eye movements were tracked and recorded. A series of self-report questions was asked to capture the participants' perceptions after each ad was played. Before watching each video, the participants' eye condition was checked using a nine-point calibration and validation exercise. Finally, eye-movement data on key objects were exported from the system and used for analysis.

Eye-tracking methodology

Owing to advancements in eye-tracking technology, modern eye-tracking equipment makes it easy to measure visual attention because it can record consumers' eye movement under natural exposure conditions, with large amounts of stimuli, high precision and at a low cost [16]. This study used an Eye Tribe Tracker to track participants' eye movements. The tracker was placed below the screen displaying the stimuli and pointed toward the user. The Eye Tribe Tracker has a sampling rate of 60 Hz and a tracking accuracy of about 0.5 to 1 degrees of visual angle. This study used Zhang et al.'s Advertisement Video Analysis System [17] to collect and analyze the participants' eye-movement data. With the system, users can track the viewer's objective attention for each object of interest in video ads.

Measures

Control variables

The consumers' demographic characteristics (e.g. gender and age) and measures (e.g. product involvement, brand familiarity and endorser familiarity) were collected as control variables. The product involvement scale ($\alpha = 0.927$), derived from Wu et al. [18], included ten 7-point semantic differential question items such as "Important"/"Unimportant". The participants were asked to rate their familiarity with the brand and endorser on a 5-point scale ranging from 1 (very unfamiliar) to 5 (very familiar). The participants were familiar with three of the six brands and endorsers. Thus we expect a variable combination of video ads to be measured in this study.

Subjective variables

Ad efficiency measures were mainly adapted from related prior studies to suit the study context. All effect items in our study were measured using a 5-point Likert scale ranging from 1 (strongly disagree) to 5 (strongly agree). The reported attention scale, developed by Novak et al. [19], was used in this experiment to measure the subjective attention for overall video ad. The scale had the following verbal labels: "Not deeply engrossed"/"Deeply engrossed"; "Not absorbed intently"/"Absorbed intently"; "My attention was not focused"/"My attention was focused"; and "I did not concentrate fully"/"I concentrated fully". The subjective recall items were mainly derived from Wu et al. [18] with some necessary modifications. The participants were asked to indicate their AR ($\alpha = 0.826$) using the following items: "I can remember most of the ad content", "This ad enhanced my impression toward the product", "I can describe the ad content" and "When I see similar ads, I can recall this ad". We assessed ATA ($\alpha = 0.907$) using the following items: "Overall, this ad is. . . (1) 'attractive', (2) 'useful', (3) 'entertaining' and (4) 'good'" [20–21].

Objective variables

Three categories were created as objective data: product, brand and endorser. The product typology contained any area of interests (AOIs) related to the product being advertised, such as the product packaging; the brand typology contained any AOIs related to the brand name (could be graphical or textual in nature) or the brand's logo; and the endorser typology contained any AOIs that advertised the endorser, whether human or animated.

Furthermore, for each key element three eye-tracking metrics are defined and presented: transformed fixation time (TFT, %), transformed fixation number (TFN, %) and average gaze duration (AGD). The fixation time and fixation number are the most commonly used indexes measures of attention on AOIs in existing literature [22]. Recent research has shown that the number and duration of fixations can explain 45% of the variance in the actual in-market sales performance of television commercials [8]. As each dynamic AOI appeared for varying lengths of time in the video ads, it was important to standardize the fixation time and fixation number to minimize the influence of the AOI showing time. Therefore, the fixation time and fixation number on the AOIs were converted into a standardized percentage. Divided by the AOI showing lengths, fixation time and fixation number were converted to TFT and TFN variables, respectively. As the third index, the AGD was calculated by dividing the participants' fixation time by their fixation number on the AOI.

Regression results

To explore the role of attention on each element in regard to their associations with consumers' reported attention, memory and attitude to product-based video advertising, we constructed three separate logistic regression models to address the research questions, controlling for the effect of gender and age by entering them as covariates into the model. In addition, variables such as product involvement, brand familiarity and endorser familiarity were controlled to compare attention predictors' relative impact on the dependent variables (DVs).

Logistic regression analyses showed that in reported attention (RAT) model, gender, age, brand familiarity and AGD of product for product objects were positive predictors of reported attention to the ad, while AGD for endorser objects was a negative predictor for reported attention (Table 3.1). In ad recall (AR) model, gender, brand familiarity and AGD for product objects were all positive related to consumers' recall for ad (Table 3.2). In attitude toward ad (ATA) regression model, brand familiarity, endorser familiarity and AGD on the product objects were all positive predictors of consumers' attitude toward ad (Table 3.3).

Discussion

Research questions

This study reveals that users' subjective reported attention of the overall video ad is related to objective attention of product objects (AGD) and endorser objects (AGD). A longer AGD on the product objects is expected to produce a greater RAT, while a lower AGD on endorser tends to produce a greater reported attention effective. It implies a correspondence between subjective and objective measures, which may reflect participants' strategy of scanning the video ad.

Users' subjective recall of ad is associated with product elements only. The longer the user's AGD on the product objects, the better their recall performance, which is likely because consumers' memory of the video ad is goal oriented. As Anderson and Pichert [23] reported, the recall of information might be biased by the goal at retrieval. That is, memory depends mainly on the product elements, which are the core of product-focused video advertising. When being exposed to product-focused video advertising, consumers automatically transform the attention of product into memory. However, objective attention to the brand and the endorser might not be processed into memory. Thus the recall of ad information might be mainly associated with product attention.

Users' ATA are associated with the product objects (AGD) in this study. The longer the users' AGD on the product objects, the better their attitudes towards the ad, which might be biased by the goal. This finding is consistent with Keller's [24] report that consumers' processing goals during ad exposure also affects their evaluations.

Table 3.1 Forward stepwise logistic regression analysis to predict the reported attention (RAT) vs. participant demographic (gender, age), product involvement, familiarity to brand and endorser, and eye-movement attention to product, brand and endorser

Variables in the Final Model	B	SE	Wald	OR	95% C.I.	
					Lower	Upper
Step 1						
AGD of product	1.594	.272	34.385	4.926***	2.891	8.393
−2 LL	460.231					
Nagelkerke R^2	.155					
Model χ^2	45.304***					
df	1					
Step 2						
Brand familiarity	.277	.072	14.990	1.319***	1.147	1.518
AGD of product	1.328	.273	23.613	3.772***	2.208	6.443
−2 LL	445.012					
Nagelkerke R^2	.204					
Model χ^2	60.523***					
df	2					
Step 3						
Brand familiarity	.292	.073	16.029	1.339	1.161	1.545
AGD of product	1.435	.285	25.334	4.198***	2.401	7.339
AGD of endorser	−1.873	.658	8.115	.154**	.042	.557
−2 LL	436.570					
Nagelkerke R^2	.229					
Model χ^2	68.966***					
df	3					
Step 4						
Age	.063	.025	6.498	1.065*	1.015	1.118
Brand familiarity	.281	.074	14.516	1.324***	1.146	1.530
AGD of product	1.473	.287	26.310	4.364***	2.485	7.662
AGD of endorser	−1.991	.670	8.840	.137**	.037	.507
−2 LL	429.691					
Nagelkerke R^2	.250					
Model χ^2	75.845***					
df	4					
Step 5						
Gender[a]	.499	.240	4.328	1.648*	1.029	2.637
Age	.057	.025	5.407	1.059*	1.009	1.112
Brand familiarity	.287	.074	14.858	1.332***	1.151	1.541
AGD of product	1.503	.289	26.995	4.497***	2.550	7.928
AGD of endorser	−2.181	.685	10.137	.113**	.030	.432
−2 LL	425.315					
Nagelkerke R^2	.263					
Model χ^2	80.220***					
df	5					

Notes: AGD = Average gaze duration; * $p < .05$. ** $p < .01$. *** $p < .001$. [a] Gender (1) = Males.

Table 3.2 Forward stepwise logistic regression analysis to predict the ad recalls (AR) effectiveness vs. participant demographic (gender, age), product involvement, familiarity with brand and endorser, and eye-movement attention to product, brand and endorser

Variables in the Final Model	B	SE	Wald	OR	95% C.I.	
					Lower	Upper
Step 1						
AGD of product	1.739	.284	37.376***	5.691	3.259	9.939
−2 LL	455.169					
Nagelkerke R^2	.174					
Model χ^2	51.122***					
df	1					
Step 2						
Brand familiarity	.244	.072	11.505**	1.276	1.108	1.469
AGD of product	1.491	.286	27.255***	4.443	2.538	7.777
−2 LL	443.560					
Nagelkerke R^2	.210					
Model χ^2	62.730***					
df	2					
Step 3						
Gender[a]	.541	.232	5.457*	1.718	1.091	2.706
Brand familiarity	.248	.072	11.731**	1.282	1.112	1.478
AGD of product	1.521	.287	28.045***	4.576	2.606	8.033
−2 LL	438.035					
Nagelkerke R^2	.227					
Model χ^2	68.255***					
df	3					

Notes: AGD = Average gaze duration; $^*p < .05.$ $^{**}p < .01.$ $^{***}p < .001.$ [a] Gender (1) = Males.

Eye metrics

This finding implies that differences exist between eye-movement metrics AGD, TFT and TFN in terms of reported attention, memory and affection effectiveness toward video advertising. The AGD on the product was a good metric to predict RAT, AR and ATA, which indicate that one eye fixation with a long gaze duration, instead of a number of short fixations, improves the memory performance and consumer attitude for dynamic information on video advertising. Furthermore, AGD may be an important eye indicator for dynamic objects and it is closely related to the core objects in the video ad.

Conclusion

This paper examined the effectiveness of video ads in terms of how the objective attention, measured by eye movement, to key ad elements was associated with video ad outcomes. By measuring and analyzing users' eye-movement data using three eye-movement metrics, we found that advertising effectiveness, reported

Table 3.3 Forward stepwise logistic regression analysis to predict the attitude toward ad (ATA) vs. participant demographic (gender, age), product involvement, familiarity with brand and endorser, and eye-movement attention to product, brand and endorser

Variables in the Final Model	B	SE	Wald	OR	95% C.I.	
					Lower	Upper
Step 1						
AGD of product	2.403	.314	58.577***	11.062	5.977	20.470
−2 LL	385.258					
Nagelkerke R^2	.296					
Model χ^2	88.549***					
df	1					
Step 2						
Endorser familiarity	.335	.082	16.904***	1.398	1.192	1.640
AGD of product	1.907	.327	34.031***	6.736	3.549	12.785
−2 LL	368.436					
Nagelkerke R^2	.345					
Model χ^2	105.371***					
df	2					
Step 3						
Brand familiarity	.235	.082	8.139**	1.264	1.076	1.485
Endorser familiarity	.259	.086	9.164**	1.296	1.096	1.533
AGD of product	1.814	.326	30.933***	6.138	3.238	11.633
−2 LL	360.432					
Nagelkerke R^2	.367					
Model χ^2	113.375***					
df	3					

Notes: AGD = Average gaze duration; * $p < .05$. ** $p < .01$. *** $p < .001$.

attention, ad recall and attitude toward ad have positive association with the eye movement of product elements through the AGD matrix. In addition, subject attention was negatively related to eye-movement measure of endorser elements through AGD. In summary, participants' eye-movement data to the elements when looking at the video ads throws light on relations between the three design elements and traditional ad effect constructs: attention, recall and attitude. The eye-movement measure AGD revealed to be a powerful indicator for predicting the product's objective attention to have effects on cognitive processing and attitudinal changes.

This study not only enriches our understanding of the relation between eye movement to elements and video ad effectiveness but also extends our knowledge of eye-movement metrics for tracking dynamic objects. These results have important implications for designing video advertising that enable customers to engage in the cognitive and affection processing to foster purchaser behavior. The results also help enrich the marketing theory and improve the marketing success model, demonstrating the potential of eye-tracking capabilities for decision-making in advertising business intelligence.

Authors' note

This work was supported by the fund of MOE (Ministry of Education in China) Project of Humanities and Social Sciences (Grant No. 17YJC880076), Educational Commission of Fujian Province, China (Grant No. JAS170171).

References

[1]   Alexander, R., Dias, S., Hancock, K.S., Leung, E.Y., Macrae, D., Ng, A.Y., . . . & Westberg, T. (2001). US Patent No. 6,177,931. Washington, DC: US Patent and Trademark Office.

[2]   Wedel, M., & Pieters, R. (2008). Eye tracking for visual marketing. *Foundations and Trends® in Marketing*, 1(4), 231–320.

[3]   Griffith, D.A., Krampf, R. F., & Palmer, J.W. (2001). The role of interface in electronic commerce: Consumer involvement with print versus on-line catalogs. *International Journal of Electronic Commerce*, 5(4), 135–153.

[4]   Poels, K., & Dewitte, S. (2006). How to capture the heart? Reviewing 20 years of emotion measurement in advertising. *Journal of Advertising Research*, 46(1), 18–37.

[5]   Duchowski, A. (2007). *Eye tracking methodology: Theory and practice (Vol. 373)*. Springer Science & Business Media.

[6]   Egeth, H. E., & Yantis, S. (1997). Visual attention: Control, representation, and time course. *Annual Review of Psychology*, 48(1), 269–297.

[7]   Milosavljevic, M., & Cerf, M. (2008). First attention then intention: Insights from computational neuroscience of vision. *International Journal of Advertising*, 27(3), 381–398.

[8]   Venkatraman, V., Dimoka, A., Pavlou, P.A., Vo, K., Hampton, W., Bollinger, B., . . . & Winer, R. S. (2015). Predicting advertising success beyond traditional measures: new insights from neurophysiological methods and market response modeling. *Journal of Marketing Research*, 52(4), 436–452.

[9]   Niederdeppe, J. (2005). Assessing the validity of confirmed ad recall measures for public health communication campaign evaluation. *Journal of Health Communication*, 10(7), 635–650.

[10] Wooley, B. (2015). The influence of dynamic content on visual attention during television commercials (Doctoral dissertation, Murdoch University).

[11] Meyers-Levy, J., & Sternthal, B. (1991). Gender differences in the use of message cues and judgments. *Journal of Marketing Research*, 84–96.

[12] Zaichkowsky, J.L. (1985). Measuring the involvement construct. *Journal of Consumer Research*, 12(3), 341–352.

[13] Kent, R. J., & Allen, C.T. (1994). Competitive interference effects in consumer memory for advertising: The role of brand familiarity. *The Journal of Marketing*, 97–105.

[14] Faul, F., Erdfelder, E., Lang, A. G., & Buchner, A. (2007). G* Power 3: A flexible statistical power analysis program for the social, behavioral, and biomedical sciences. *Behavior Research Methods*, 39(2), 175–191.

[15] Ooms, K., Dupont, L., Lapon, L., & Popelka, S. (2015). Accuracy and precision of fixation locations recorded with the low-cost Eye Tribe tracker in different experimental setups. *Journal of Eye Movement Research*, 8(1).

[16] Poole, A., & Ball, L. J. (2006). Eye tracking in HCI and usability research. *Encyclopedia of Human Computer Interaction*, 1, 211–219.

[17] Zhang, X. B., Fan, C. T., Yuan, S. M., & Peng, Z. Y. (2015, December). An Advertisement video analysis system based on eyetTracking. In Smart City/SocialCom/SustainCom (SmartCity), 2015 IEEE International Conference on (pp. 494–499). IEEE.

[18] Wu, S. I., Wei, P. L., & Chen, J. H. (2008). Influential factors and relational structure of internet banner advertising in the tourism industry. *Tourism Management*, 29(2), 221–236.

[19] Novak, T. P., Hoffman, D. L., & Yung, Y. F. (2000). Measuring the customer experience in online environments: A structural modeling approach. *Marketing Science*, 19(1), 22–42.

[20] Pieters, R., Wedel, M., & Batra, R. (2010). The stopping power of advertising: Measures and effects of visual complexity. *Journal of Marketing*, 74(5), 48–60.

[21] Goodrich, K., Schiller, S. Z., & Galletta, D. (2015). Consumer reactions to intrusiveness of online-video advertisements. *Journal of Advertising Research*, 55(1), 37–50.

[22] Jacob, R. J., & Karn, K. S. (2003). Eye tracking in human-computer interaction and usability research: Ready to deliver the promises. *Mind*, 2(3), 4.

[23] Anderson, R. C., & Pichert, J. W. (1978). Recall of previously unrecallable information following a shift in perspective. *Journal of Verbal Learning and Verbal Behavior*, 17(1), 1–12.

[24] Keller, K. L. (1987). Memory factors in advertising: The effect of advertising retrieval cues on brand evaluations. *Journal of Consumer Research*, 14(3), 316–333.

4 An optimization algorithm and smart model for periodic capacitated arc routing problem considering mobile disposal sites

Erfan Babaee Tirkolaee and Ali Asghar Rahmani Hosseinabadi

Introduction

Today, generating different types of solid wastes and emersion of their related social, economic and environmental incompatibilities create numerous complications for urban service management regarding the collection, transportation, processing and disposal of such wastes. Due to the fact that 60%–80% of costs of managing solid wastes are dedicated to the collection and transportation (Tirkolaee et al., 2018a), the assessment of these collection and transportation systems plays an important role in reducing and solving the problems of urban service management and create an efficient financial plan. In other words, generating a decision support system in financial domains of these organizations will be the best way. Nowadays, big data can be used for strategic policymaking in almost any field of the urban waste management (Lu et al., 2016).

In an ideal system, wastes should be collected, transported and disposed in the least time using the best possible method, and directly from the residential sites to the disposal site. Based on the above discussions, the importance of establishing an optimal system for waste collection becomes more highlighted. Therefore, planning the optimal policy of waste collection plays a significant role in reducing costs. In terms of the routing of urban waste collection, two categories of problems are defined. In the first category, there is a predefined node series and the objective is to find the best routes that pass across all these nodes. In the second category, a series of edges are defined in the network and the objective is to find the best routes passing through all the required edges (edges with positive demands). The collection of urban wastes is included in the second category.

Regarding the problem of collecting household wastes, they are distributed along the edges (i.e. streets or alleys with demands). In addition, the capacity of vehicles is limited and they can collect only a given volume of wastes. As to the problem of locating and routing periodical arches with mobile disposal sites described in the present paper, the days of the week are divided into even and odd days. Mobile disposal sites are considered as the requirements of modern real-life applications. The objective is to determine the optimal routes of serving the required edges based on the categorization of days, defining the location and number of mobile disposal sites and restarting the serving operation by the vehicles.

In the following, the associated literature is reviewed briefly. As to periodical models, Pia and Filippi (2006) presented a new version of capacitated arc routing problem (CARP) for two types of vehicles, in which the first type of vehicle could unload at the depot while the second type unloads its wastes into the vehicles of the first type (Beltrami and Bodin, 1974). In their proposed model, determination of optimal routes for each vehicle is incorporated in optimal decision-making for two types of vehicles. As a result, their problem is specific to collecting domestic wastes. They solved this problem through the variable neighborhood descent (VND) algorithm (Solomon, 1987).

Chu et al. (2005) introduced a periodic capacitated arc routing problem (PCARP) for programming a weekly horizon. Their problem model was based on integer linear programming and they solved their problem through two innovative methods. The objective of their problem was to assign a set of service days to each edge in a defined network and solve the resulting routing arch for each period to minimize the size of the necessary fleet and total costs of the trip during the predefined time horizon.

This problem has numerous applications for street networks such as collection of wastes and clearing away snow. Lacomme et al. (2005) developed a PCARP for practical applications such as collecting waste. They describe few models of PCARP with a simple categorization scheme. For instance, demand for each wing might be dependent upon the programming period or the last date of serving. They used an evolution algorithm (EA) based on a complex intersection operator to simultaneously modify and apply operational and tactical decision-making such as service days and daily trips. Ogwueleka (2009) suggested an innovative method for solving the problem of collecting waste in Onitsha, Nigeria. The comparison results with current situation showed that a vehicle can be successfully excluded from service. In addition, the length of the route, collection costs and collection time respectively decrease to 16.31%, 25.24% and 23.51%.

Babaee Tirkolaee et al. (2016) proposed a novel mathematical model for the robust CARP. The objective of their proposed model was to minimize total traversed distance considering the demand uncertainty of the required edges. They could design a hybrid meta-heuristic algorithm based on **Simulated annealing (SA)** algorithm and a heuristic algorithm in order to solve the problem efficiently.

Tirkolaee et al. (2018a) presented a multi-trip CARP for urban waste collection in order to minimize total cost. In their proposed problem, depots and disposal facilities were located in different places according to the real-world condition. They developed a hybrid algorithm method based on an Improved Max-Min Ant System (IMMAS) to solve some well-known benchmark problems and large-sized instances.

Tirkolaee et al. (2018b) proposed a novel mathematical model for the robust PCARP considering working time of the vehicles. They designed a hybrid SA algorithm in order to solve the large-sized problem approximately. They concluded that their proposed algorithm could generate proper robust solutions in comparison with the exact method.

Tirkolaee et al. (2018c) developed a hybrid **Genetic Algorithm (GA)** for solving a multi-trip green CARP in the field of urban waste collection. They presented a novel model incorporating a CARP model for waste collection and environmental aspects of the problem in transportation. They could generate appropriate applicable solutions in a reasonable run time.

The technologies addressed in the present paper includes consideration of usage time limitation of vehicles to determine the number of the required vehicles, consideration of mobile disposal site, possible of numerous trips of a vehicle based on time limits, developing a mathematical model to simultaneously consider the location and period of each problem, developing an innovative method to generate the primary solution and introducing two algorithms of simulated annealing.

The following sections discuss the problem and its mathematical model, introduce the suggested methodology and develop the numerical results of implementing the algorithm. The final section provides the conclusion and further suggestions.

Problem description and smart model

In the present paper, the problem is the determination of an optimal number of vehicles and their optimal tours based on minimizing total objective function which includes the cost of using vehicles, the cost of passing across network edges and the cost of operationalizing mobile sites to unload the vehicles. In this regard, the vehicles at the depot (Node 1) start their travel to serve required edges. After finishing their capacity, the vehicles move towards the desired disposal site ($D = \{n, n+1, \ldots, n+d\}$, D: Set of potential nodes) to recover their initial capacity, restart their trip from disposal site in the case of sufficient time and return to the operational location. In addition to considering the limitation of capacity, time bounds of each vehicle are similarly significant. When the remaining time for a vehicle tends towards zero, it has to return to the disposal site. After recovering initial loading capacity (i.e. disposal), they go back to their depot.

- The objective function of the developed model should include minimization of costs of passing across the edges, minimization of fleet size and the number of necessary vehicles to satisfy the across the edges, minimization of fleet size and the number of required vehicles to satisfy the total demand as well as the cost of using the mobile site.
- Each required edge is served only by one vehicle.
- One depot and a few mobile disposal sites exist. One of the objectives of the problem in the present study is to determine their location.
- Vehicles start their trip from the depot and when their capacity is full, they return to the disposal site and initiate their trip in the case of sufficient remaining time.
- The vehicles are heterogenous and the graph network is asymmetrical.

- There is no failure of the collection. This means that each edge should be served once and by one vehicle.
- After ending the trip, the vehicles return to the depot.
- Each vehicle has a maximum duration of service.
- The time and costs of passing through an identical route by different vehicles are similar.
- The days of the week are divided into odd and even ones. Number "1" represents even days and number "2" represents odd days ($t = \{1, 2\}$).

This model with the above definition can be considered a decision support system (DSS) in order to create an optimal planning. Sets, indexes, parameters, decision variables and the mathematical model are described in the following:

Sets:

V	Set of all nodes
S	Set of mobile site nodes
E	Set of all defined edges
E_R	Set of all required edges
E_A	Set of required edges that must supply daily

Indices:

i, j	Nodes in V set
S	Nodes in S set
k	Vehicle
t	Period

Parameters:

K	Maximum number of vehicles
T	Total number of period
Pk	P-th tour of k-th vehicle
tt_{ij}	Time of passing the edge (i, j)
c_{ij}	Cost of passing the edge (i, j)
d_{ij}	Demand of edge (i, j)
cv_k	Cost of using k-th vehicle
FC_s	Fixed cost of establishing and operating the s-th mobile site
T_{max}	Maximum available time of each vehicle
W_k	Capacity of k-th vehicle
M	Large number

Decision variables:

x^t_{ijpk}	If edge (i, j) is traveled n times by k-th vehicle during p-th tour in period t, variable is n, otherwise 0.
y^t_{ijpk}	If edge (i, j) is supplied by k-th vehicle during p-th tour in period t, variable is 1, otherwise 0.
YY_{st}	If s-th mobile site is used in period t, variable is 1, otherwise 0.
u_k	If k-th vehicle is used is 1, otherwise 0.

Mathematical model:

$$Minimize\ Z = \sum_{(i,j)\in E}\sum_{k\in K}c_{ij}x^t_{ij\,p_k} + \sum_{k\in K}cv_k u_k + \sum_{s\in S}\sum_{t=1}^{2}FC_s YY_{st} \tag{4.1}$$

$$\sum_{j=1}^{n}x^t_{ij\,p_k} = \sum_{j=1}^{n}x^t_{jip_k},\forall i\in V\setminus\{1,S\},\forall k\in K,\forall p_k$$
$$=1,2,\ldots,P_k,\forall t\in\{1,2\} \tag{4.2}$$

$$\sum_{p_k\in P_k}\sum_{t=1}^{2}\left(y^t_{ijp_k}+y^t_{jip_k}\right)=1,\forall(i,j)or(j,i)\in E_R \tag{4.3}$$

$$\sum_{p_k\in P_k}\left(y^t_{ijp_k}+y^t_{jip_k}\right)=1,\forall(i,j)or(j,i)\in E_A,\forall t\in\{1,2\} \tag{4.4}$$

$$\sum_{(i,j)\in E_R}q_{ij}y^t_{ijp_k}\le W_k,\forall k\in K,\forall p_k=1,2,\ldots,P_k,\forall t\in\{1,2\} \tag{4.5}$$

$$y^t_{ijp_k}\le x^t_{ijp_k},\forall(i,j)\in E,\forall k\in K,\forall p_k=1,\ldots,P_k,\forall t\in\{1,2\} \tag{4.6}$$

$$\sum_{p_k\in P_k}\sum_{t=1}^{2}\sum_{(i,j)\in E}x^t_{ijp_k}\le Mu_k,\forall k\in K \tag{4.7}$$

$$\sum_{p_k\in P_k}\sum_{(i,j)\in E}tt_{ij}x^t_{ijp_k}\le T_{max},\forall k\in K,\forall t\in\{1,2\} \tag{4.8}$$

$$\sum_{i=2}^{V-S}\sum_{p_k\in P_k}x^t_{idp_k}\le MYY_{st},\forall k\in K,\forall s\in S \tag{4.9}$$

$$\sum_{(i,j)\in E_R}\sum_{p_k\in P_k}\sum_{k\in K}d_{ij}y^t_{ijp_k}\le Cap_s YY_{st},\forall t\in\{1,2\},\forall s\in S \tag{4.10}$$

$$\sum_{j=2}^{V-S}x^t_{1j\,p_k}=u_k YY_{st},\forall k\in K,\forall p_k=1,\forall t\in\{1,2\},\forall s\in S \tag{4.11}$$

$$\sum_{j=2}^{V-S}x^t_{j\,d\,p_k}=u_k YY_{st},\forall k\in K,\forall p_k=1,\forall t\in\{1,2\},\forall s\in S \tag{4.12}$$

$$\sum_{j=2}^{V-S}x^t_{d\,j\,p_k}\le u_k YY_{st},\forall k\in K,\forall p_k=2,\ldots,P_k,\forall t\in\{1,2\},\forall s\in S \tag{4.13}$$

$$\sum_{j=2}^{V-S}x^t_{j\,d\,p_k}\le u_k YY_{st},\forall k\in K,\forall p_k=2,\ldots,P_k,\forall t\in\{1,2\},\forall s\in S \tag{4.14}$$

$$\sum_{i,j\in R}x^t_{i\,j\,p_k}\le1+nh^R_{p_k},\forall R\subseteq V\setminus\{1,S\};R\ne\varnothing;\forall k\in K,$$
$$\forall p_k=1,\ldots,P_k,\forall t\in\{1,2\} \tag{4.15}$$

$$\sum_{i \in R} \sum_{j \neq R} x^t_{ijp_k} \geq 1 - f^R_{p_k}, \forall R \subseteq V \setminus \{1, S\}; R \neq \varnothing; \forall k \in K,$$

$$\forall p_k = 1, \ldots, P_K, \forall t \in \{1, 2\} \tag{4.16}$$

$$h^R_{p_k} + f^R_{p_k} \leq 1, \forall R \subseteq V \setminus \{1, S\}; R \neq \varnothing; \forall k \in K, \forall p_k = 1, \ldots, P_K \tag{4.17}$$

$$x^t_{ijp_k} \in Z^+, y^t_{ijp_k} \in \{0, 1\}, u_k \in \{0, 1\}, h^R_{p_k} \in \{0, 1\}, f^R_{p_k} \in \{0, 1\}, \tag{4.18}$$

$$\forall (i, j) \in E, \forall k \in K, \forall t \in \{1, 2\}, \forall k \in K, \forall p_k = 1, \ldots, P_K,$$

$$\forall R \subseteq V \setminus \{1, S\}; R \neq \varnothing$$

The objective function 4.1 has three parts. The first part includes minimizing the costs of pasting from (i, j) edge by k-th vehicle, the second part consists of minimizing the cost of buying (hiring) the k-th vehicle and the third part includes the costs of establishing and operating the s-th mobile site in period t. As shown in limitations (equation 4.2), the flow equilibrium relations for each vehicle are developed. The limitations in equation 4.5 represent the limitation of capacity of the k-th edge. As shown in equation 4.6, the limitations show that the required edge is served by a traversing vehicle (i.e. a vehicle might pass across an edge without serving it). The denoted limitations in equation 4.7 imply that the k-th vehicle is used during period t and its costs are paid. The denoted limitations in equation 4.8 show the time limit of each vehicle. The limitations in equation 4.9 denote that the s-th mobile site is used in period t, the cost of which is paid out. The limitations in equation 4.10 represent the capacity limit of the s-th mobile site in period t. The limits in equations 4.11 and 4.12 guarantee that the first trip of the vehicle starts from the depot and the rest of the trips initiate from the disposal site. The limitations in equation 4.13 guarantee that all vehicle trips end at the disposal site. The denoted limitations in equations 4.14–4.17 ascertain that there is no non-authorized tour. As to limitations denoted in equation 4.18, types of the variables are defined.

Considering multiple trips for each tour

To consider the possibility of a difference of depot from the disposal site as well as numerous trips in each tour, a series of trips with index P_k is considered. The first trip of each vehicle starts from the depot (equation 4.13) and after traveling across some edges, it returns to the disposal site (equation 4.15). For the next trip, the vehicle starts its travel from the disposal site (equation 4.14), and after going across some edges it returns to the disposal location.

Optimization algorithm

In the literature, there have been many solution methods developed in order to solve the research optimization problem and the other similar optimization problems (Alinaghian et al., 2014; Babaee Tirkolaee et al., 2016; Mirmohammadi et al., 2017, Tirkolaee et al., 2017; Yuan et al., 2017, Metawa et al., 2017; Hosseinabadi et al., 2018a; Hassanien et al., 2018; Hassan et al., 2018; Elhoseny et al., 2015; Elhoseny et al., 2015; Hosseinabadi et al., 2017; Hosseinabadi et al., 2018b; Elhoseny et al., 2018; Mostafaeipour et al., 2018; Hosseinabadi and Tirkolaee, 2018). Due to the high complexity of the problem, solving the model is inefficient through precise methods. Therefore, an innovative algorithm is suggested for developing initial solutions and two algorithms of simulated annealing are developed. In regard to the latter, the algorithms are different in terms of cooling mechanism and number of iterations. In the next sections, the mechanism of the innovative algorithm to provide initial solutions and develop an SA algorithm is detailed as an intelligent information system (IIS). IIS can be realized by providing online real-world data, especially demand of edges.

Constructive algorithm to develop initial solution

To develop the initial solution, an innovative random algorithm is used, the execution steps of which are detailed in the following:

Step 1: Randomly choose one period among the existing periods and go to step 2.

Step 2: Randomly choose one vehicle among the existing vehicles and go to step 3.

Step 3: Randomly choose one disposal site among the existing disposal site and go to step 4.

Step 4: Start a trip from the depot and go to step 5.

Step 5: Among the edges the starting point of which is from the depot, randomly select one edge among k-edges with the least duration from the depot and go to step 6.

Step 6: If there is an edge attached to the current node (node at the end of the current edge), go to step (6). Otherwise, use the solution of the shortest path problem recursively and choose the shortest middle edge(s). Then, go to step 7.

Step 7: In the case of attachment of edges to the current node, consider k edges of the highest level of demand for which the capacity and time limits of the vehicle enables one to select them. Randomly select one of the vehicles and go to step 8. The edge should cover the time limit which enables one to pass from it and go to the disposal location in a definite period of time. Otherwise, check the following two conditions:

    A: If the remaining time does not allow a traversing/serving for a vehicle and there is still some time, refer to the predefined disposal location and go to step 1. If there is no period left, select the period with the least number of vehicles (in identical conditions, randomly select one) and go to step 2.

    B: If the capacity limit is not considered, go to the defined site, zero the capacity and go to step 7.

Step 8: Add the resulting demand of waste collection in the selected edge to the vehicle and deduct the demand of the edge from the list of demands. Check the following condition. If all required edges are crossed, end the algorithm. Otherwise, go to step (7).

The number of developed solutions is intended to be 200. When the algorithm ends, the best solution is used for the SA algorithm.

Representation of solution

To display the solution, a string of variable length is employed in which each tour of the vehicle in a period of time is considered to be a vector. In other words, the string only covers justifiable tours which include all necessary edges. To better understand the defined string, Figure 4.1 shows an instance which includes six required edges and two vehicles in two periods of odd and even days. The strings represented in Table 4.1 include the solutions with three vectors. To interpret the solution, the first vector is assigned based on time limitation of the first period, the second vector is assigned due to time bounds of the first vehicle and the second vehicle is assigned to the first period. The third vector represents the assignment of the second vehicle to the first period.

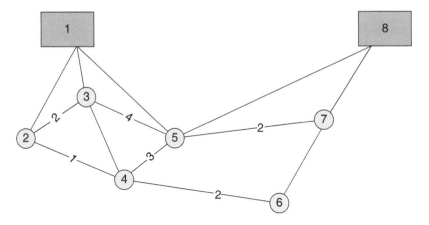

Figure 4.1 An instance of the developed string

Table 4.1 Representing solution

First period of first vehicle	1 2 3 4 5 8 5 7 8
First period of second vehicle	1 2 4 6 7 8
Second period of first vehicle	1 3 4 5 8

Improving solution through algorithm of simulated annealing

To improve the solutions, two SA algorithms are used and all initial solutions are distinctively improved by the algorithm. The SA algorithm is a meta-heuristic algorithm of local search which can refrain from local optimization. It is a very efficient algorithm for solving problems with non-convex or discrete solution space. As a result, SA is used to solve an integer programming problem (Glover and Kochenberger, 2005). In addition, ease of implementation, homogeneity and hill climbing to refrain from local optimization make SA a proper choice for improving the generated initial solutions after each iteration. The basis of this algorithm starts with an initial algorithm solution followed by consideration of a neighborhood it. If the value of the objective function of the neighbor is better than the objective function of the solution, the solution is substituted by its neighbor. Otherwise, a number is randomly made within a range of zero and one and it is compared with the value defined by the algorithm. If the random number is less than that of the algorithm, the worse solution is accepted. The condition for ending the algorithm is to attain the final temperature. Before starting the algorithm, the following items should be defined:

1   Cooling mechanism (i.e. the number of algorithm iterations in each temperature (M), number of changes in temperature, mechanism/ relation of reducing temperature)
2   Initial temperature
3   Final temperature.

First simulated annealing algorithm (SA1)

In this algorithm, the following equation is used to reduce the temperature:

$$t_{k+1} = \alpha t_k \; ; \; 0 < \alpha < 1 \tag{4.19}$$

In the above equation, t_k represents the value of temperature in the k-th iteration. Of course, in the associated literature, the range of $0.8 < \alpha < 0.99$ is recognized for assuring a relatively slow and efficient scheduling. It is evident that when x is higher, the rate of reduction of temperature decreases and it will be possible to search the problem space more.

Number of SA1 iterations in each temperature

In this algorithm, the counter m is used to count the accepted solution for each iteration. This counter should attain the extent of parameter M in each iteration.

Second simulated annealing algorithm

In this algorithm, a slower test is done through the following cooling mechanism to reduce energy.

$$t_{k+1} = \lim_{\beta \to 0} \frac{t_k}{1 + \beta t_k} \tag{4.20}$$

In this equation, β is a small optional value.

Number of SA2 iterations in each temperature

In SA2, the number of iterations for each temperature is obtained from the following equation:

$$M\left(t_k\right) = \frac{t_k}{1+t_k} \tag{4.21}$$

In the present paper, due to the randomized selection of three problems as instances, a trial-and-error method is used to define the values of algorithm parameters, as shown in Table 4.2.

Local search

To develop a completely random solution with the adjacency of the present solution, the following algorithm is used:

Step 1: Randomly choose a row of current solution and call it *s*.

Step 2: Randomly choose a node from the first row and call it *R*.

Step 3: Randomly choose a row as the second row and call it *u*. If all rows have been tested before, go to step 2.

Step 4: Look for node *R* in row *u*. If you did not find it, go to step 3.

Step 5: From the start of the path to *R* in row *s* and from *R* to the end of the path in row *u* of the current solution should be placed in row *s*.

Step 6: From the start of the path to *R* in row *u* and from *R* to the end of the path in row *s* of the current solution should be placed in row *u*.

Step 7: Add the utilized site in row *s* from the current solution to row *u* of the new solution.

Step 8: Add the utilized site in row *u* from the current solution to row *s* of the new solution (due to the intersection of these two rows, the violated limitation of using one site for each row can be satisfied through steps 7 and 8).

Step 9: Transfer the remaining rows from the current solution to the new one.

In this method, all limitations of the plausibility of the new solution have been satisfied with the exception of time bounds of the tours and limitation of loading capacity of trips. The limitations of capacity are satisfied through considering the intended objective function. This leads to a diversity of solutions to the problem

Table 4.2 Values of algorithm parameters

Parameters	Values
M	5
α	0.98
β	0.0001
t	200
t	1
η	0.1

and increased probability to attain the proper solution. To depict this mechanism, see Figure 4.2. The third and fourth rows are selected for developing neighborhood and the intended node is node 3. The solution in the neighborhood of current solution is developed in the following manner.

Computational results

To investigate the efficiency of the proposed smart model and in order to validate it, ten instance problems are generated and the qualities of the obtained solutions are verified. The information of these problems is shown in Table 4.3. It is noteworthy that the problems are examined in two periods.

In Table 4.3, the first column represents the problems, the second column refers to total number of the network nodes, the third column denotes the total number of edges, the fourth column represents the total number of required edges, the fifth column refers to the number of potential mobile sites, and the last column denotes the number of available vehicles for each problem.

To evaluate the efficiency of SA algorithm for desired problems, the results of the algorithms are compared with those of solver CPLEX. The SA1 and SA2 algorithms were coded in the programming language C#. To execute the application, a 2.5 GHz CPU and 4 GB of RAM is used. Each problem is solved through meta-heuristic algorithms and its best value is regarded as proper. A summary of comparative results is shown in Table 4.4.

As shown in Table 4.4, the mean error values for SA1 and SA2 algorithms compared with CPLEX are 3.78% and 2.27%, respectively, which denoted the higher

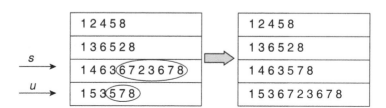

Figure 4.2 Developing a solution in the neighborhood

Table 4.3 Input information

Problem	TN	TE	RE	TDS	AV
P1	7	8	6	2	2
P2	10	20	16	2	2
P3	12	35	28	3	2
P4	14	50	33	3	2
P5	15	60	49	3	2

Table 4.4 Computational results

Problem	CPLEX		SA1			SA2		
	Optimal Value	Run Times (sec)	Best Value	Run Time (sec)	GAP Percentage	Best Value	Run Time (Sec)	GAP Percentage
P1	458.11	15.52	472.63	17.3	3.17	461.09	22.3	0.65
P2	662.51	30.54	693.98	20.1	4.75	682.85	27.02	3.07
P3	871.36	53.72	905.95	24.09	3.97	891.84	32.2	2.35
P4	980.39	145.33	1003.82	30.3	2.39	995.19	35.1	1.51
P5	1493.1	333.68	1543.42	35.17	3.37	1521.62	41.38	1.91
P6	1526.89	808.43	1586.44	47.88	3.9	1591.17	56.25	4.21
P7	1941.19	1174.6	2019.81	53.41	4.05	1985.26	57.86	2.27
P8	2218.45	1645.18	2267.26	61.24	2.2	2243.07	69.28	1.11
P9	2541.27	3600	2684.60	70.4	5.64	2608.36	81.5	2.64
P10	3284.15	3600	3284.15	83.64	0	3284.15	90.87	0
Average	1597.74	1140.70	1646.21	44.35	3.34	1626.46	51.38	1.97

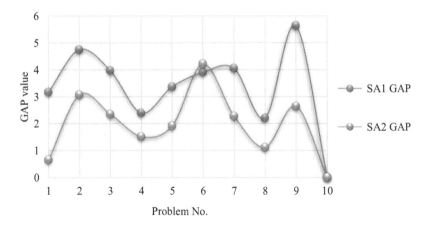

Figure 4.3 Comparison of the performance of SA1 and SA2

efficiency of SA2 in regard to finding proper solutions. To exemplify this issue more clearly, see Figure 4.3.

Regarding run times of the solution methods, as shown in Figure 4.4, the mean time of solving random instances in the precise method is almost 905.36 seconds, while for SA1 and SA2 algorithms, the necessary duration for solving the developed problems are 44.35 and 51.37 seconds, respectively, which are almost identical. The obtained information represents the proper speed of suggested algorithms.

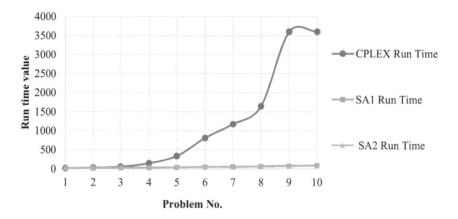

Figure 4.4 Run time comparison of CPLEX and SA algorithms

Conclusion and further suggestions

The routing and optimal assignment of vehicles for problems of routing vehicles is one of the significant financial planning of organizations such as municipalities in regard to collecting urban wastes because the proper assignment of vehicles and optimal routing can reduce a large portion of associated costs.

In the present problem, a mixed integer linear programming smart model is used in order to consider a periodic capacitated arc routing problem with mobile disposal sites specified for the collection of urban wastes. The obtained solutions contain the following items:

1   The vehicle that can be used
2   The day in which edges are served
3   The sites or mobile sites used on odd and even days
4   The optimal routes passed during the days of the week.

So, this system is considered as a decision support system in order to be implemented by the municipalities. Optimization algorithms of two types of simulated annealing are used in order to solve the problem. The results obtained in the small-sized problems showed that these algorithms generate appropriate solutions in comparison with CPLEX. It has been concluded that the proposed algorithm can implement an intelligent information system through online real-world data to generate online solutions.

As future suggestions on getting the model close to real-world ones, internet of things application, big data analytics for customer value creation, and customer segmentation or profiling can be considered in the problem.

References

Alinaghian, M., Amanipour, H., & Tirkolaee, E. B. (2014). Enhancement of inventory management approaches in vehicle routing-cross docking problems. *Journal of Supply Chain Management Systems*, 3(3).

Babaee Tirkolaee, E., Alinaghian, M., Bakhshi Sasi, M., & Seyyed Esfahani, M. M. (2016). Solving a robust capacitated arc routing problem using a hybrid simulated annealing algorithm: A waste collection application. *Journal of Industrial Engineering and Management Studies*, 3(1), 61–76.

Beltrami, E., & Bodin, L. D. (1974). Networks and vehicle routing for municipal waste collection. *Networks*, 4, 65–94.

Chu, F., Labadi, N. & Prins, C. (2005). Heuristics for the periodic capacitated arc routing problem. *Journal of Intelligent Manufacturing*, 16(2), 243–251.

Elhoseny, M., Yuan, X., Yu, Z., Mao, C., El-Minir, H. K., & Riad, A. M. (2015). Balancing energy consumption in heterogeneous wireless sensor networks using genetic algorithm. *IEEE Communications Letters*, 19(12), 2194–2197.

Elhoseny, M., Tharwat, A., & Hassanien, A. E. (2018). Bezier curve based path planning in a dynamic field using modified genetic algorithm. *Journal of Computational Science*, 25, 339–350.

Glover, F. W., & Kochenberger, G. (2005). *Handbook of meta-heuristics*. Norwell: Kluwer Academic Publishers.

Hassan, M. K., El Desouky, A. I., Badawy, M. M., Sarhan, A. M., Elhoseny, M., & Gunasekaran, M. (2018). EoT-driven hybrid ambient assisted living framework with naïve Bayes – firefly algorithm. *Neural Computing and Applications*, 1–26.

Hassanien, A. E., Rizk-Allah, R. M., & Elhoseny, M. (2018). A hybrid crow search algorithm based on rough searching scheme for solving engineering optimization problems. *Journal of Ambient Intelligence and Humanized Computing*, 1–25.

Hosseinabadi, A.A.R., Rostami, N.S.H., Kardgar, M., Mirkamali, S., & Abraham, A. (2017). A new efficient approach for solving the capacitated vehicle routing problem using the gravitational emulation local search algorithm. *Applied Mathematical Modelling*, 49, 663–679.

Hosseinabadi, A.A.R., Vahidi, J., Saemi, B., Sangaiah, A. K., & Elhoseny, M. (2018a). Extended genetic algorithm for solving open-shop scheduling problem. *Soft Computing*, 1–18.

Hosseinabadi, A.A.R., Vahidi, J., Balas, V. E., & Mirkamali, S. S. (2018b). OVRP_GELS: solving open vehicle routing problem using the gravitational emulation local search algorithm. *Neural Computing and Applications*, 29(10), 955–968.

Hosseinabadi, A.A.R., & Tirkolaee, E. B. (2018). A gravitational emulation local search algorithm for task scheduling in multi-agent system. *IJOAS*, 1(12), 11–24.

Lacomme, C., & Prins, W. R. (2005). Evolutionary algorithms for periodic arc routing problems, *European Journal of Operational Research*, 165(2), 535–553.

Lu, W., Chen, X., Ho, D. C., & Wang, H. (2016). Analysis of the construction waste management performance in Hong Kong: the public and private sectors compared using big data. *Journal of Cleaner Production*, 112, 521–531.

Metawa, N., Hassan, M. K., & Elhoseny, M. (2017). Genetic algorithm based model for optimizing bank lending decisions. *Expert Systems with Applications*, 80, 75–82.

Mostafaeipour, A., Qolipour, M., Rezaei, M., & Babaee-Tirkolaee, E. (2018). Investigation of off-grid photovoltaic systems for a reverse osmosis desalination system: A case study. Desalination.

Mirmohammadi, S.H., Babaee Tirkolaee, E., Goli, A., & Dehnavi-Arani, S. (2017). The periodic green vehicle routing problem with considering of the time-dependent urban traffic and time windows. *Iran University of Science & Technology*, 7(1), 143–156.

Ogwueleka, T. Ch. (2009). Municipal solid waste characteristics and management in Nigeria. *Iranian Journal of Environmental Health Science & Engineering*, 6(3), 173–180.

Pia, A., & Filippi, C. (2006). A variable neighborhood descent algorithm for a real waste collection problem with mobile depots. *International Transactions in Operational Research*, 13(2), 125–141.

Solomon, M, M. (1987). Algorithms for the vehicle routing and scheduling problem with time window constraints. *Operations Research*, 35(2), 254–265.

Tirkolaee, E.B., Goli, A., Bakshi, M., & Mahdavi, I. (2017). A robust multi-trip vehicle routing problem of perishable products with intermediate depots and time windows. *Numerical Algebra, Control & Optimization*, 7(4), 417–433.

Tirkolaee, E.B., Hosseinabadi, A.A.R., Soltani, M., Sangaiah, A.K., & Wang, J. (2018a). A hybrid genetic algorithm for multi-trip green capacitated arc routing problem in the scope of urban services. *Sustainability*, 10(5), 1–21.

Tirkolaee, E.B., Alinaghian, A., Hosseinabadi, A.A.R., Sasi, M.B., & Sangaiah, A.K. (2018b). An improved ant colony optimization for the multi-trip capacitated arc routing Problem. *Computers & Electrical Engineering*, in press.

Tirkolaee, E.B., Mahdavi, I., & Esfahani, M.M.S. (2018c). A robust periodic capacitated arc routing problem for urban waste collection considering drivers and crew's working time. *Waste Management*, 76, 138–146.

Yuan, X., Elhoseny, M., El-Minir, H.K., & Riad, A.M. (2017). A genetic algorithm-based, dynamic clustering method towards improved WSN longevity. *Journal of Network and Systems Management*, 25(1), 21–46.

5 Opinion mining analysis of e-commerce sites using fuzzy clustering with whale optimization techniques

K. Shankar, M. Ilayaraja, P. Deepalakshmi,
S. Ramkumar, K. Sathesh Kumar, S. K.
Lakshmanaprabu and Andino Maseleno

Introduction

The expression "web-based social networking" includes an extensive variety of online exercises, blogs, company exchange sheets, chats, service rating sites, micro-web journals and so forth [1]. In the present world, e-business goals are getting the opportunity to be capital of market. No one needs to go outside for [2] advertising because of the constancy of this site and these destinations are more trusted than features [3]. Web shopping sections empower customers to buy things by methods for the web and moreover get them passed on to wish lists for an apparent charge, in this way diminishing the time required to buy a thing [4]. Supposition mining takes in individuals' viewpoints, tests, feelings toward people, individuals, issues, activities, subjects and their features [5]. The opinion is broad because they are fundamental effects of our practices [6]. Presently, numerous individuals were taking about the movies and the subject of the movies in various aspects [7].

Feature extraction recognizes those points of view which are being commented by customers, suspicion gauge recognizes the substance containing supposition or appraisal by picking thought limit as positive, negative or impartial in the conclusion rundown module [8]. The component is evaluated by the customers explicitly, using the exactness as an appraisal metric to affirm the features extraction and examination process [9]. These days, delicate registering techniques are vivaciously passed on in the e-trade business as data warehousing, and "delicate figuring" is the center of data warehousing or some other move advancement [10]. The subjective points of view are an aggregation of feelings, reviews, recommendations, comments, assessments and individual experiences shared by different customers compiled through get-togethers, casual associations, and close by the credible data [11].

Literature review

Singh V. and Dubey S.K. (2014) summarized the sentiment analysis of social events. The sentiment analysis is very broad and consists of natural language processing, text mining, decision-making and linguistics. The researchers focus on the product-based classification, but factors such as analysis of sentiment, emotions and opinion are the main factors for product classification. The information from the social sites, blogs and micro blogging sites are used for the investigation of individuals' feelings and opinion [12].

The opinion mining using the K means clustering and Markov Random Field (MRF) feature selection is proposed for Thai restaurant owners to improve their business. In the text processing, the reviews are separated into words, stop words are removed and then texts are transformed for creating keywords and generating input vectors. The MRF feature selection reduces the number of features in the dataset and K means clustering is employed for the best clustering performance [13].

In 2017, Aditya Parashar et al. [14] built up a basic leadership calculation to assess the nature of an item by arranging past audits on a size of numbers and showing it over an e-trade site. Clients can utilize these positions over any e-commerce site to settle on their own decisions.

In 2016, Seyedali Mirjalili et al. [15] proposed that the calculation is motivated by the air pocket net chasing technique. The Whale Optimization Algorithm (WOA) is tried with 29 scientific improvement issues and six auxiliary plan issues. Advancement comes about to demonstrate that the WOA calculation is extremely focused, contrasted with the condition-of-workmanship meta-heuristic calculations and also ordinary strategies.

Park S. and Kim D.Y. (2017) have proposed the online recommender systems within the context of the tourism industry to estimate the language discrepancies. The traveler preferences are captured through analyzing textual data. Text clustering and Jaccard distance score text mining techniques are used to examine the language discrepancies. The adequacy of tourism sites is evaluated by contrasting the idea of the dialect travelers used to portray the experience and substance given by the goal advertisers on tourism sites [16].

Methodology for ranking model

In our paper (Figure 5.1), the new application situated model proposed online items in various shopping sites in view of the audits of customers. At first, we create the extremity database of various items distinctive sources from e-trade locales. Every item has its own features sets are great markers in clustering the item audits in light of above chosen features (attributes). Initially user reviews are collected from existing customers of each product. The whole client surveys refreshed component based outline is created for gathering using the FCM model. In the wake of collection examination consider the WOA optimization to improve the features to discover ideal sites with items, Moreover this investigation positioning procedure to rank the ideal features based suggestion e-trade locales with risk priority number (RPN) investigation [17].

Collection of reviews

We prepared crude dataset from online e-shopping sites (Amazon, Flipkart, snapdeal, Paytm, etc.). Every one of these locales is extremely prominent in introduce from where the greater part of the general population gets a kick out of the chance

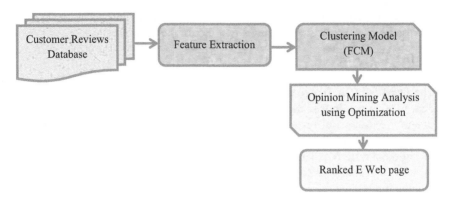

Figure 5.1 Block diagram for proposed model

to buy. A large portion of the item includes are things and the greater part of the words used to decide the extremity of these features are descriptive words found in the region of the opinion feature.

Opinion mining analysis

In all actuality, associations and affiliations reliably need to find buyer or general appraisals about their things and organizations. The machine learning technique can empower us to know cases of the notions. For the most part this examination has three stages, including feature extraction, grouping and raking model; this features related to that particular component taken in to thought, build up in light of those features the entire execution capacities are attempted and genuine and moreover includes system to choose a fresh or various course of action of classes, the new social occasions are of stress in themselves, and their valuation is regular.

Feature extraction from reviews

The feature extraction and process is involved to separate the features from the collected item reviews to enhance the quality of the review for exami-nation. The connection amongst suppositions and item includes enhances the item audit rating. The example attributes data like cost, positive surveys, qual-ity and so forth; each includes in every last audit to demonstrate the client articulations.

Clustering topology

Since clustering of the client opinions is helpful to different perspectives of busi-ness and furthermore, it is the common strategy used to find many element articu-lations from content for a feeling mining application. This paper proposes the social event of reviews from online goals using cushioned bundling technique. To recognizing the clustering audits of human and dissents and seeing those with a sort one can require the demand grouping. Grouping is an unsupervised learning task, so no class regards addressing a past mix of the data cases.

Fuzzy clustering

The FCM is by and large used for clustering where the execution of the FCM depends on upon the assurance of beginning bunch focus or participation incentive to the features of reviews. It gives a strategy for how to assemble information focuses that populate some multidimensional space into a particular number of various clustering. The fundamental target of iterative clustering and fuzzy c-Means calculation is to limit the weight inside clustering entirety of squared blunder target capacity and it appeared beneath the equation.

$$Objective = \sum_{i=1}^{d}\sum_{j=1}^{c} r_{ij}^{e} \parallel l_i - c_j \parallel^2 \qquad (5.1)$$

In the above equation the parameters objective function and fuzziness index, d, l, c as membership of ith data to jth clustering center, feature vector and jth clustering center. The FCM enables each element vector to have a place with each bunch with a fluffy truth esteem (in the vicinity of 0 and 1), which is figured utilizing equation 5.1. Fuzzy clustering by differentiating permits information focuses to have a place within excess of one gathering. Each clustering is related to an enrollment work that communicates how much individual information directs have a place in the group.

$$c_j = \left(\frac{\sum_{i=1}^{d}(l_{ij})^e f_i}{\sum_{i=1}^{d}(l_{ij})^e} \right) \qquad (5.2)$$

The related clustering focuses that speak to the structure of the information as most ideal as the calculation depends on the client to indicate the number of groups introduces in the arrangement of information to be grouped. At last, these means with the exception of the underlying advance are rehashed until the centroids never again move. From the technique of fuzzy clustering process, the chose data of items are groups a few clustering like cost, quality, shipping charge and a few parameters.

Optimization technique

The mathematical optimization method, which is a best method to choose an element from a group of obtainable alternatives, is used in mathematics, computer science and operation research [18]–[28]. Simply, an optimization issue contains a set of maximum or minimum real functions from which selecting an input value from an acceptable value and calculating the value of the function [29]–[35]. Mostly, the optimization theory and methods are used in the field of applied mathematics. The optimization method also includes finding the best accessible value of target function from a defined domain or variety of target functions from different type of domain [36]–[42].

Optimization ranking analysis

Features determination and clustering based positioning the thing surveys breathed life into one straightforward streamlining is considered. The proposed system positions the thing picked by the client, not for all things. As indicated by that special groups find the positioning rating for a specific thing with information. By then, proposed structure glance through the thing with assurance showed by the customer. This situating in light of the base cost, maximum quality, optimal brand and its objective work showed as:

$$Ranking = \{\min\_ \cos t, \max\_ quality, opt\_ brand\} \tag{5.3}$$

To establish client trust, e-business people should set base criteria for the item if the item reviews are coming into this range than they ought to be considered the further deal on their stage. The decision of ranking plan relies upon the purchaser's inclination. The refresh position for featured groups appeared beneath the section.

Whale optimization

In the meta-heuristic algorithm, a recently proposed optimization called WO, which motivated from the bubble-net hunting strategy. The QO algorithm is based on the extraordinary hunting behavior of humpback whales. The whales produces bubbles in a circular or 9-shaped path. It will encircle the prey and follows the bubbles and moves upward the surface. This entire procedure is used for feature selection and is shown below.

WHALES INITIALIZATION

Initialize the population of particles (here, the features from clustering) is defined in terms of D_i

$$F_{ei} = \{F_{e1}, F_{E2}, F_{E3}, \ldots \ldots F_{en}\} \tag{5.4}$$

ENCIRCLING PREYS

The target prey and the other hunt operators try to revive their positions towards it and this behavior showed up by the accompanying condition.

$$P = |K.F*(t) - F(t)| \tag{5.5}$$

$$F_{(t+1)} = F*(t)_{best} - O.*P \tag{5.6}$$

From equations 5.5 and 5.6 $O = 2.r$ where F^* best position, F as whale position.

BUBBLE-NET ATTACKING METHOD

Two enhanced methodologies are outlined as follows for numerically simulating the bubble-net behavior of humpback whales.

Spiral-updating position   The separation between the position of whale and its prey is ascertained, and afterward an equation of spiral is made amongst whale and prey areas to simulate the development of helix shape by humpback whales.

$$F(t+1) = R.e^{bt}.\cos(2\pi.a) + F^*(t) \tag{5.7}$$

Where a is random value between -1 and 1, humpback whales swim around the prey inside a contracting circle and along a spiral-shaped path simultaneously. So we expect that there is a possibility around 50% to choose between either the shrinking encircling mechanisms.

SEARCH FOR PREY   Successively to have an overall analyzer, the query operator is revived by a randomly selected search agent rather than the best search agent.

$$R = \left| \vec{C}.F_{rand} - F \right| \tag{5.8}$$

$$F(t+1) = \vec{F}_{rand} - AK\,R \tag{5.9}$$

This mechanism and $|K| > 1$ accentuate exploration and permit the WO algorithm to play out a global optimum and $|K| >$ for refreshing the position of the search agents.

TERMINATION CRITERIA

The procedure is repeated until the point when the greatest number of iterations is to come. For updating position of the artificial whales in the scan space and for reenacting the developments of whales, two vectors, specifically, step vector ($\Delta Feature$) and position ($Feature$) are considered.

Reviews ranking

After choosing the optimal ranking opinions to assess the relative significance of each element as per related sentiment score and utilize that measure to rank features. Also, we trust that these commentators rating of sites are the vital data that advances distinguish untruthful opinions. They at that point propose a perspective positioning calculation to rank the vital angles by thinking about both viewpoint recurrence and impact of opinions given to every perspective on their general sentiments.

Results analysis

The proposed framework takes the dataset from the arbitrary web-based shopping site where the audits about the icebox, cell phones et cetera. This strategy executed in JAVA platform windows machine containing arrangements Intel (R) Core i5 processor with 1.6 GHz and 4 GB RAM. The example items with the site the surveys appeared in Table 5.1.

Reviews with evaluations in this range represent almost 80% of aggregate surveys, mirroring a for the most part great involvement with the items in the Amazon. Table 5.1 demonstrates some most mainstream sites with real items. Review scores are broken down. Database showcasing, through client database data, organizations can dissect the purchaser inclinations of clients, and give clients distinguishing electronic lists to build the attraction list to customers.

Table 5.2 demonstrates the diverse items vs. sites score regard examination, In a huge segment of the suppositions, per clients use to state concerning some book that substance of the book is immaterial or book seems to club everything together, this sort of irrelevancy is orchestrated as a substitute segment and it is given weight in negative an impetus to anticipate the rank in more correct and exact measures. By then ordinary on precision and typical on audit is processed for every feeling.

Figure 5.2 demonstrates the all techniques bunched the reviews into different groups each clustering incorporate positive or negative comments. The trial after-effects of fuzzy clustering contrasted and K means clustering in view of features. This proposed clustering model precision at 89.62% for the audit information 256 features, while contrasted with another model. That figure demonstrates the exactness and review of various sites in the wake of clustering process, general altogether improved measurements on in our proposed model.

Table 5.1 Number of reviews data

Websites/ products	Electronic application	Mobile phone	Books	Fashion
Amazon	258	105	128	452
Flipkart	342	471	94	236
Snapdeal	278	453	115	248
Paytm	21	198	123	489

Table 5.2 Score values of products with features for proposed model

Products	Amazon	Flipkart	Snapdeal	Paytm
P1 – Electronic application	10	6.25	9.45	4
P2 – Mobile Phone	10	8.55	9.2	8
P3 – Books	8	8	8.55	7.2
P4 – Fashion	9.8	9	6.8	7.5

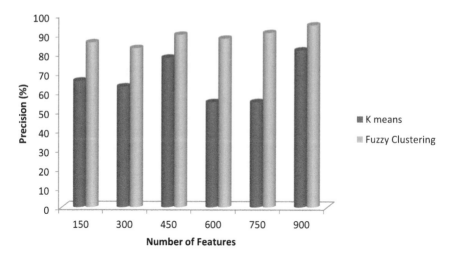

Figure 5.2 Result of feature clustering

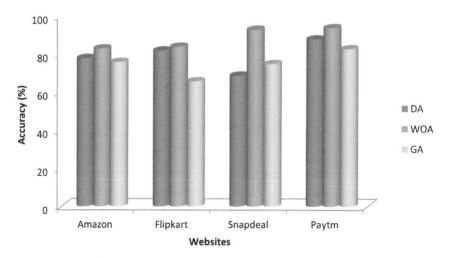

Figure 5.3 Optimization results

Figure 5.3 demonstrates the positioning rating limit of the reviews, and age of the survey and the supportiveness thing score of the audits for registering the score for a thing. Our proposed advancement display thought about hereditary calculation (GA), dragonfly algorithm (DA). The general thing figuring is done and the best assessed things appears to the customer with thing examination with

WOA change and without improvement are dissected, among this examination the most extraordinary score rating addressed in the Amazon site.

To analyses the proposal framework, different kinds of suggestion algorithms are utilized. A few items in internet business were taken in view of assumptions. At that point normal on exactness and normal on review is figured for each opinion. The result appears (Figure 5.4) that proposed structure gave the better outcome in examination of system. Sensitivity, specificity and precision, F measure is low in the FCM examination when contrasted with the FCM with DA methods.

With a specific end goal to assess the feasibility of sentiment sifting, we welcomed ten volunteers to physically rate the audits that have been sifted through. Figure 5.5 appears about the changed class of item category, similar to cost, quality, delivery days, shipping charges and brand. The greatest best score achieved on Amazon site for greatest cases, so the vast majority of the clients like this site just for internet shopping reason.

Conclusion

As shown by our approach to demonstrate the most effective web-based shopping site and it's carrying on. The execution of the item in the wake of checking the item execution whether great or poor in light of client surveys from internet business website. Unhelpful surveys can be sifted through naturally from all the buyer audits with a high review rate around 90% and 94% accuracy. It is sincerely necessary for an organization to know the assessments of clients about its items. The positioning of items, item score, the correlation between in excess of two items, prescribe item list alongside its general score. This paper we have to think about clustering for features and advancement likewise we have used to

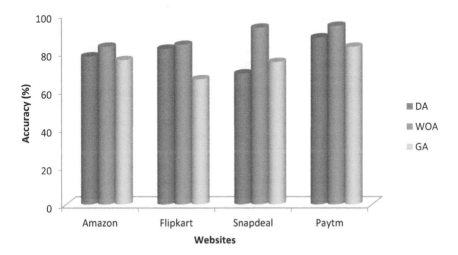

Figure 5.4 Comparative analysis

Figure 5.5 Ranking analysis

positioning reason. The creators know which features of a particular thing are required and which features ought to be upgraded to extend the customer dedication. As a future change, the structure may be contacted use portrayal strategy for aggregate positive and negative studies for better legitimization.

References

[1] Heng, Y., Gao, Z., Jiang, Y. and Chen, X. (2018). Exploring hidden factors behind online food shopping from Amazon reviews: A topic mining approach. *Journal of Retailing and Consumer Services*, 42, 161–168.

[2] Sadhana, S.A., SaiRamesh, L., Sabena, S., Ganapathy, S. and Kannan, A. (2017, February). Mining Target Opinions from Online Reviews Using Semi-supervised Word Alignment Model. In *Recent Trends and Challenges in Computational Models (ICRTCCM), 2017 Second International Conference on* (pp. 196–200). IEEE.

[3] Liao, S.H., Chen, Y. J. and Lin, Y.T. (2011). Mining customer knowledge to implement online shopping and home delivery for hypermarkets. *Expert Systems with Applications*, 38(4), 3982–3991.

[4] Babu, A.G., Kumari, S. S. and Kamakshaiah, K. (2017, January). An Experimental Analysis of Clustering Sentiments for Opinion Mining. *In Proceedings of the 2017 International Conference on Machine Learning and Soft Computing* (pp. 53–57). ACM.

[5] Kumar, K.S., Desai, J. and Majumdar, J. (2016, December). Opinion mining and sentiment analysis on online customer review. In *Computational Intelligence and Computing Research (ICCIC), 2016 IEEE International Conference on* (pp. 1–4). IEEE.

[6] Oza, K. S. and Naik, P. G. (2016). *Prediction of Online Lectures Popularity: A Text Mining Approach.* Procedia Computer Science, 2nd International Conference on Intelligent Computing, Communication & Convergence (ICCC-2016), Odisha, India.

[7] Sohail, S. S., Siddiqui, J. and Ali, R. (2016). Feature extraction and analysis of online reviews for the recommendation of books using opinion mining technique. *Perspectives in Science*, 8, 754–756.

[8] Sohail, S. S., Siddiqui, J. and Ali, R. (2016). Feature extraction and analysis of online reviews for the recommendation of books using opinion mining technique. *Perspectives in Science*, 8, 754–756.

[9] Li, G. and Liu, F. (2012). Application of a clustering method on sentiment analysis. *Journal of Information Science*, 38(2), 127–139.

[10] Wang, X., Chai, X., Hsu, C. H., Xiao, Y. and Li, Y. (2015, August). Cluster analysis based on opinion mining. In Ubi-Media Computing (UMEDIA), *2015 8th International Conference on* (pp. 110–115). IEEE.

[11] Dubey, G., Rana, A. and Shukla, N. K. (2015, February). User reviews data analysis using opinion mining on web. In *Futuristic Trends on Computational Analysis and Knowledge Management (ABLAZE), 2015 International Conference on* (pp. 603–612). IEEE.

[12] Singh, V. and Dubey, S. K. (2014, September). Opinion mining and analysis: A literature review. In *Confluence the Next Generation Information Technology Summit (Confluence), 2014 5th International Conference* (pp. 232–239). IEEE.

[13] Claypo, N. and Jaiyen, S. (2015, January). Opinion mining for Thai restaurant reviews using K-Means clustering and MRF feature selection. In *Knowledge and Smart Technology (KST), 2015 7th International Conference on* (pp. 105–108). IEEE.

[14] Parashar, A. and Gupta, E. (2017, February). ANN based ranking algorithm for products on E-Commerce website. In *Advances in Electrical, Electronics, Information, Communication and Bio-Informatics (AEEICB), 2017 Third International Conference on* (pp. 362–366). IEEE.

[15] Mirjalili, S. and Lewis, A. (2016). The whale optimization algorithm. *Advances in Engineering Software*, 95, 51–67.

[16] Park, S. and Kim, D. Y. (2017). Assessing language discrepancies between travelers and online travel recommendation systems: Application of the Jaccard distance score to web data mining. *Technological Forecasting and Social Change*, 123, 381–388.

[17] Lakshmanaprabu, S. K., Shankar, K., Gupta, D., Khanna, A., Rodrigues, J.J.P.C., Pinheiro, P. R. and de Albuquerque, Victor Hugo C. (2018). Ranking analysis for online customer reviews of products using opinion mining with clustering. *Complexity* (In press)

[18] Avudaiappan, T., Balasubramanian, R., Sundara Pandiyan, S., Saravanan, M., Lakshmanaprabu, S. K. and Shankar, K. (2018). Medical image security using dual encryption with oppositional based optimization algorithm. *Journal of Medical Systems-Springer.* September 2018. DOI: https://doi.org/10.1007/s10916-018-1053-z.

[19] Karthikeyan, K., Sunder, R., Shankar, K., Lakshmanaprabu, S. K., Vijayakumar, V., Elhoseny, M. and Manogaran, G. (2018). Energy consumption analysis of Virtual Machine migration in cloud using hybrid swarm optimization (ABC – BA), *The Journal of Supercomputing – Springer.* DOI: https://doi.org/10.1007/s11227-018-2583-3.

[20] Shankar, K., Lakshmanaprabu, S. K., Gupta, D., Maseleno, A. and de Albuquerque, V.H.C. (2018). Optimal feature-based multi-kernel SVM approach for thyroid disease classification. *The Journal of Supercomputing*, 1–16.

[21] Lydia, E.L., Kumar, P.K., Shankar, K., Lakshmanaprabu, S.K., Vidhyavathi, R. M. and Maseleno, A. (2018). Charismatic document clustering through novel K-Means Non-negative Matrix Factorization (KNMF) algorithm using key phrase extraction. *International Journal of Parallel Programming*, 1–19.

[22] Shankar, K., Elhoseny, M., Lakshmanaprabu, S. K., Ilayaraja, M., Vidhyavathi, R.M., and Alkhambashi, M. (2018). Optimal feature level fusion based ANFIS classifier for brain MRI image classification. *Concurrency and Computation: Practice and Experience – Wiley*, June 2018. DOI: https://doi.org/10.1002/cpe.4887.

[23] Lakshmanaprabu, S.K., Shankar, K., Khanna, A., Gupta, D., Rodrigues, J.J., Pinheiro, P. R. and De Albuquerque, V.H.C. (2018). Effective features to classify big data using social internet of things. *IEEE Access*, 6, 24196–24204.

[24] Shankar, K. and Eswaran, P. (2016). RGB-based secure share creation in visual cryptography using optimal elliptic curve cryptography technique. *Journal of Circuits, Systems and Computers*, 25(11), 1650138.

[25] Shankar, K. and Eswaran, P. (2015). A secure visual secret share (VSS) creation scheme in visual cryptography using elliptic curve cryptography with optimization technique. *Australian Journal of Basic & Applied Science*, 9(36), 150–163.

[26] Shankar, K. and Eswaran, P. (2015). ECC based image encryption scheme with aid of optimization technique using differential evolution algorithm. *International Journal of Applied Engineering Research*, 10(55), 1841–1845.

[27] Shankar, K. and Eswaran, P. (2016). An efficient image encryption technique based on optimized key generation in ECC using genetic algorithm. In *Artificial Intelligence and Evolutionary Computations in Engineering Systems* (pp. 705–714). Springer, New Delhi.

[28] Shankar, K. and Lakshmanaprabu, S.K. (2018). Optimal key based homomorphic encryption for color image security aid of ant lion optimization algorithm. *International Journal of Engineering & Technology*, 7(1.9), 22–27.

[29] Batle, J., Farouk, A., Naseri, M. and Elhoseny, M. (2016). New approach to finding the maximum number of mutually unbiased bases in C6. *Applied Mathematics & Information Sciences An International Journal*, 10(6), 2077–2082. http://doi.org/10.18576/amis/100609.

[30] Elhoseny, M., Elleithy, K., Elminir, H., Yuan, X. and Riad, A. (2015). Dynamic clustering of heterogeneous wireless sensor networks using a genetic algorithm towards balancing energy exhaustion. *International Journal of Scientific & Engineering Research*, 6(8), 1243–1252.

[31] Rizk-Allah, R.M., Hassanien, A.E., Elhoseny, M. and Gunasekaran, M. (2018). A new binary salp swarm algorithm: development and application for optimization tasks. *Neural Computing and Applications*, 1–23.

[32] Gharbia, R., Hassanien, A.E., El-Baz, A.H., Elhoseny, M. and Gunasekaran, M. (2018). Multi-spectral and panchromatic image fusion approach using stationary wavelet transform and swarm flower pollination optimization for remote sensing applications. *Future Generation Computer Systems*, 88, 501–511.

[33] Hassanien, A.E., Rizk-Allah, R. M. and Elhoseny, M. (2018). A hybrid crow search algorithm based on rough searching scheme for solving engineering optimization problems. *Journal of Ambient Intelligence and Humanized Computing*, 1–25.

[34] Hassan, M.K., El Desouky, A.I., Badawy, M.M., Sarhan, A.M., Elhoseny, M. and Gunasekaran, M. (2017). EoT-driven hybrid ambient assisted living framework with naïve Bayes – firefly algorithm. *Neural Computing and Applications*, 1–26.

[35] Gafar, M. G., Elhoseny, M. and Gunasekaran, M. (2018). Modeling neutrosophic variables based on particle swarm optimization and information theory measures for forest fires. *The Journal of Supercomputing*, 1–18.

[36] Tharwat, A., Gaber, T., Hassanien, A. E. and Elhoseny, M. (2018). Automated toxicity test model based on a bio-inspired technique and AdaBoost classifier. *Computers & Electrical Engineering*, 71, 346–358.

[37] Elhoseny, M., Oliva, D., Osuna-Enciso, V., Hassanien, A. E. and Gunasekaran, M. (2018). Parameter identification of two dimensional digital filters using electromagnetism optimization. *Multimedia Tools and Applications*, 1–18.

[38] Hosseinabadi, A.A.R., Vahidi, J., Saemi, B., Sangaiah, A. K. and Elhoseny, M. (2018). Extended genetic algorithm for solving open-shop scheduling problem. *Soft Computing*, 1–18.

[39] Hosseinabadi, A.A.R., Vahidi, J., Saemi, B., Sangaiah, A. K. and Elhoseny, M. (2018). Extended genetic algorithm for solving open-shop scheduling problem. *Soft Computing*, 1–18.

[40] Elhoseny, M., Tharwat, A. and Hassanien, A.E. (2018). Bezier curve based path planning in a dynamic field using modified genetic algorithm. *Journal of Computational Science*, 25, 339–350.

[41] Metawa, N., Hassan, M. K. and Elhoseny, M. (2017). Genetic algorithm based model for optimizing bank lending decisions. *Expert Systems with Applications*, 80, 75–82.

[42] Elhoseny, M., Tharwat, A., Farouk, A. and Hassanien, A.E. (2017). K-coverage model based on genetic algorithm to extend WSN lifetime. *IEEE Sensors Letters*, 1(4), 1–4.

6 Big data text mining in the financial sector

Mirjana Pejić Bach, Živko Krstić and Sanja Seljan

Introduction

The financial sector generates a vast amount of data like customer data, logs from their financial products and transaction data that can be used in order to support decision-making, together with external data like social media data and data from websites (Zhao et al., 2011). Finacle Connect (2018) indicates the top ten technologies for financial industries, including the rise of API (Application Programming Interface) economy, cloud business enablement, blockchain for banking and the usage of artificial intelligence, including data analytics, machine learning, deep learning, natural language processing and visual recognition.

Turner et al. (2012), in the executive report prepared for the IBM Institute for Business Value, indicate that 71% of banking and financial institutions use big data and analytics for creating competitive advantage for their companies. The same authors state that in 2010 there were 36% of such banking and financial institutions, indicating an increase of 97% in two years. This increase points out the relevance of big data technologies in today's business for long-standing business challenges in the banking and financial sector. Applications of big data in the financial sector are various, including social media analysis, web analytics, risk management, fraud detection and security intelligence.

The goal of the chapter is to discuss data-driven case studies from the field of text mining supported by big data technologies, with a focus on social media and website data.

Background

About big data

Big data denotes a vast and complex amount of data in structured documents (data organized into records identified by a unique key, with each record having the same number of attributes), semi-structured documents (documents having structure, but differentiating among themselves, e.g. XML documents, HTML files) and unstructured documents (in terms of record layout, embedded metadata or collected textual data). Mathew from Oracle Financial Services (2012) points out several issues in big data analytics: diversification of data types along with vast amount of data, more changes and uncertainty, more unanticipated questions and real-time needs and decision-making.

Big data can be described from the traditional 3-V view, including volume, variety and velocity (Furht and Villanustre, 2016). Volume indicates vast amounts of data coming from different sources (e.g. social media, internal systems, networks, logs and sensors) in real time. Variety indicates diversity of sources and formats, while velocity indicates speed of data ingestion into big data system. Since companies aim to extract value from big data, one more V, indicating value, is suitable for the definition of big data from the business point of view. Value indicates the search for valuable insight in big data that can help companies to lower costs, increase revenue and optimize processes by analyzing large volumes of data.

Other authors include even more Vs, like variability and veracity (e.g. Gandomi and Haider, 2015; Saju and Shaja, 2017). Big data technologies support the process of gathering, grouping and examining data (Saju and Shaja, 2017). Alexander et al. (2017) indicate steps of cleaning, transformation, integration, modelling and analytics to find hidden useful information such as trends, relationships, preferences and other useful information. Therefore, big data technologies are used to aggregate large amounts of data that come to our big data system from different sources, which is important in information extraction and decision-making.

Big data architectures

Various big data architectures are used in practice, including Lambda, Kappa, SMACK and Liquid architectures (Marz and Warren, 2015). The big data architectures that are the most often used for big data analytics in financial sector are Lambda and Kappa.

The Lambda architecture (Marz and Warren, 2015) is architecture that consists of three layers: batch, speed and serving. The batch layer is part of architecture where raw data or processed data can be stored. Arbitrary views are stored in the serving layer. The batch layer contains all data to recent hours. Because the batch layer is time-consuming, views stored in the serving layer for additional use do not contain fresh and new data (from the last hours depending on time of computation of the batch layer). New and fresh data is ingested with the support of the speed layer.

The speed layer is used for real-time computations, transformations and arbitrary views for the serving layer and is used as complement to the batch layer. Real-time usually refers to milliseconds or a few seconds of delay. Together with the batch and speed layers, one unique view on the data in the system is possible.

The serving layer generates indexes for fast queries on our arbitrary views.

With Lambda architecture, two layers (batch and speed) need to be maintained in order to create one unique view of data (in the serving layer). This problem can be partially addressed with Apache Spark, which has both a batch and speed layer under one framework. Another option is to use different architecture: Kappa.

Kappa architecture focuses only on one layer, a speed (stream) layer instead of two layers like in Lambda (batch and speed). Here, the serving layer and its views are created based on one layer, while in Lambda two layers (batch and speed layer) need to be unified.

Kappa architecture demands that each event is processed, transformed and enriched in the speed layer as data arrive in our big data system (Kreps, 2014). This is an advantage, but it also generates additional complexity in cases when duplicates occur and there is a need to cross-reference different real-time events. The serving layer can be any in-memory or persistent database, used for text analysis, databases or for full-text search.

Text-mining usage for big data systems

Business intelligence is the umbrella concept including tools, architecture, resources, databases, applications and methodologies used to analyze data important for decision-making (i.e. for policymaking).

Trends for business intelligence in the financial and banking sector are among early adopters of big data technologies, showing the main trends of the financial sector (Alexander et al., 2017; Alvarado et al., 2015; Moro et al., 2015; Srivastava and Gopalkrishnan, 2015; Turner et al., 2012; Zhao, 2013):

- Customer analytics – through customer detection, retention, relationship management and targeting (customer segmentation and profiling, enhancing customer engagement, retention and loyalty, detecting customer needs, quality analytics, sentiment analysis, best offer, through gamification, chatbots, etc.).
- Development of new business models (use of historical data for predictive models, forecasts and trading impacts, real-time view of data providing competitive advantage comparing to other financial institutions, new communication channels supporting e-banking, mobile payments), marketing, product cross-selling and pricing.
- Operational optimizations based on the ability to collect data, aggregation and integration of a variety of data (reports, transactions, email, logs, social media, free-form texts, external data, geospatial, audio, images, sensors, etc.), spam detection, high capacity of warehouse, architecture (e.g. Hadoop, NoSQL), data quality management, e-banking, mobile payments, analytic software and strong analytic capabilities.
- Risk and fraud management (e.g. mortgage fraud, money laundering, risk detection, credit fraud, crime management, confidential information leaking and mail spamming).

Electronic documents have become the primary means of storing and retrieving written communication. Data mining is focused on extracting useful knowledge (e.g. trends, patterns) from unstructured or semi-structured files, databases and XML files in order to create data-driven models such as classification, regression analysis or clustering (Moro et al., 2015).

Text mining is a particular type of data mining focused on handling unstructured or semi-structured text documents. Text mining can be used to identify research trends by identifying relevant words and relationships in order to categorize or draw conclusions (He et al., 2013; Banerjee., 2016). Fan et al. (2006) defined text mining as finding novel, previously unknown information using automatic detection from various text sources, especially elaborating on information extraction, topic tracking, summarization, clustering, categorization, concept linkage, information visualization and question answering.

From the business point of view, the goal of text mining is to listen to customers, their needs, issues, suggestions and opinions and to increase their satisfaction. Mak et al. (2011) indicated that data mining model in financial institutions could improve their performance (e.g. improve processes workflow and deepen understanding of investment behaviour). The Financial Stability Board (2017) indicates that machine learning and artificial intelligence are being rapidly adopted in the financial industry, especially to assess the credit quality, price and market insurance contracts and client interaction.

Various applications of text-mining approaches are developed, and several examples will be presented here. Kloptchenko et al. (2004) performed a research on telecommunications companies' performance data. In the paper, annual reports are used to perform clustering of financial data using self-organizing maps (SOM), followed by text analysis data by extracting meaning from the textual part of financial reports, which contain richer information in order to extract information about future financial performance by using data mining techniques (text clustering, text analysis). Mak et al. (2011) conducted clustering analysis and Association Rules (ARs) in order to gain deep understanding of customer purchasing behaviour. They aimed to discover target groups and interesting relationships among customers. They concluded that better customer segmentation could help identify customers and have marketing implications.

Especially in time of crisis, when financial institutions are faced with sustainability and demanding customers, who seek low-risk investments and products relevant for their needs, financial institutions re-evaluate market situations and customer relations management, searching for hidden information relevant for the decision-making process and business workflow. However, due to the constantly augmenting amount of documents, it is impossible to process all information, manage, organize and relate it. Therefore, a massive amount of data and documentation requires text-mining methods to process information on the specific topic.

New business intelligence technologies help institutions from the financial sector to differentiate from their competitors (Bholat et al., 2015). However, this differentiation is accomplished with a better decision-making process due to new insight from big data analysis, which is the focus of our next section.

Case study of big data and text mining in the financial sector

In order to present the background for the case study, we shall present one hypothetical big data system, for which we shall define the type of data sources, the goal of the big data system and its purpose. This system belongs to one bank, which aims to compare its competitive position on the market. The usual approach to attaining this goal would be through the market research conducted on a sample of citizens who use bank services (e.g. Davison et al., 1989). However, big data allowed that this goal could be also attained through the utilization of the abundant data available on social media and websites. Therefore, the presumption is that the bank decided to utilize big data in order to gain information of its position at the market compared to competitors.

Two main data sources are used: external and internal data sources. Internal data can be transaction data, log data or application data. External data can be from any social media and website. The goal of our big data system is to collect all relevant textual data about our hypothetical institution and its competitors. These collected data can be further enriched and analyzed to extract valuable insights that we can use for different business purposes like customer segmentation, improvements of marketing campaigns and analysis of voice of customers. While financial institutions already have some inputs from their customers in some form, the benefit of big data technologies is to compare our institution with competitors using the same objective metric (usually a machine learning–generated metric).

Examples of social media sources are Facebook and Twitter. An example of a website source is a top ten website about financial news. Since each source has its own limitations and unique elements, the first step would be to create an adequate connector for each source, which is used in order to ingest data from social media to our system thanks to big data technologies. Each connector needs to be built based on some configuration inputs in order to gather needed information. Facebook and Twitter have their own APIs, and they can provide specific information to big data systems as long as the goal is to collect data from public pages and groups (Facebook) or public accounts (Twitter). Information such as the gender of the person who wrote a comment is not available, so models are often used in order to predict this information in order to use it in further analysis.

Since the bank's main goal is to gather comments, the list of bank's competitors is created. For that purpose, the configuration parameter is used for each connector, with the list of keywords (e.g. bank1, bank2, bank3, #bank1, . . .). The big data system will use Facebook and Twitter. Searching of Facebook data requires also the search space (list of pages or groups). Two lists are sent to the Facebook API: a list of keywords and a list of pages or groups. In this example, the list of keywords related to the list of major competitor banks (e.g. bank1, bank2, bank3, #bank1, . . .) and list of pages and groups has same structure (e.g. pageid125, pageid554, groupid124, . . .). There are additional parameters like language filter. We have to stress again that only public pages and groups could be included in analysis.

Requirements for Twitter are more simple, since for this media only the list of keywords is required, and the Twitter API will return any tweets containing selcected keywords in real time.

After these parameters are entered properly in the big data architecture, the custom data connectors can start feeding the bank's big data system with real-time or near real-time data. Facebook will bring data in mini-batches and Twitter will be pure real-time feed. Those datasets contain interesting information, such as username, time stamp of the comment, web page name where the comment was posted (Facebook specific), the number of likes/favourites on that comment, the number of shares or retweets, the language detected in that comment and the keyword that was used for relevant comment detection.

Websites are more complex. Each website has its own design and layout of comments and articles. Some websites are more complex, while others are more simple, which indicates the need to build a custom connector (if website has API), to use a web crawl process (if it is possible by terms and conditions) or to use a commercial partner that can provide that information legally. A crawl process is commonly used. For crawling, a crawl connector or process and workflow for each individual website should be developed. Parameters would be the website domain, the crawl start URL (if only one category on that website is searched, e.g. financial category) and keywords that are to be found. Based on these inputs, the specific link extraction process is created for each website as well as the specific content crawl process for extracting articles and comments.

The crawling process will start based on the selected parameters and then all relevant URLs will be filtered for content crawl process. When the full list of relevant URLs is available (e.g. URLs that are related to financial topics), the needed information can be crawled, such as article, author, time stamp of article, all comments of that article, comment author, comment time stamp, and so forth.

The result of the presented big data system are datasets extracted from Facebook, Twitter and top ten websites from external sources and transnational data, log files from internal sources in order to gain metrics like number of accounts opened, number of transactions, usage from our website and so forth.

However, in order to be useful, this information should be transformed, cleaned and enriched in a big data environment. The enrichment process is related to the following text-mining case studies:

1   Keyword extraction
2   Named entity recognition
3   Gender prediction
4   Sentiment analysis
5   Topic extraction
6   Social network analysis.

Text-mining case studies

Keyword extraction

With new technologies and analysis in recent times, and especially in case of big data analytics with vast volumes of new data coming from different sources, there is a need for keyword extraction. Keyword extraction plays an important role in the financial sector. It was used in simple form in the previous example when a list of keywords was needed in order to extract related comments and articles from an external source. More complex, sophisticated usage would be to use automatic keyword extraction (Hasan and Ng, 2014). This field gained huge interest in the past several years, since volumes of data are growing and every document or comment cannot be read sequentially. The goal is to extract a "sequence of words", called n-grams, through semi-automated process. However, this process does require manual validation and comparison with a reference model (i.e. "gold standard") in order to assess the quality of the tool. Quality of terminology has gained importance regarding costs, user perceptions and customer satisfaction. For these reasons, various metrics are used in order to estimate the quality of automatically extracted terminology (Seljan et al., 2013; 2014; 2017).

In some cases, pre-processing techniques are needed, starting from the corpus collection, where documents can be collected from one or more sources depending on various criteria, including corpus size and domain. A possible step could be de-duplication of documents or articles, pre-formatting, possibly scanned and converted by OCR (optical character recognition). Once in the appropriate format, the text is tokenized, and represented as a list of words, numbers, signs and punctuation and treated as a "bag of words".

To summarize, keyword extraction has four approaches (Bharti et al., 2017): statistical (term frequency, inverse document frequency); linguistic (WordNet, n-gram, POS patterns); machine learning (naïve Bayes); and hybrid approach (some combination of the previous three approaches). Seljan et al. (2009a, 2009b, 2013, 2014) conducted analysis using a statistically based extraction of terminology and collocations from a previously aligned bilingual corpus, followed by linguistic filtering using local grammars for two languages but on corpora differing in size, domain and provenance. Seljan et al. (2013) performed research on the automatic extraction process, after the digitization process and OCR, from monolingual text by three statistically based language independent methods based on statistical occurrence with/without stop word list, c-value or term hood combined with frequency, which can be embedded into longer candidate term and followed by linguistic filtering. The aim of the research was to extract meaningful information in order to facilitate document classification. Dunđer et al. (2015) analyzed various statistically based terminology extraction approaches for the purpose of classification of archival records. While most evaluations refer to precision, recall and F-measure, or comparison with reference lists or already existing resources, Gašpar and Seljan (in print) performed evaluation of extracted terminology by Herfindahl-Hirschman Index (HHI), commonly used to measure market concentration to determine market competitiveness, but here adapted to measure terminology concentration.

A simple example used in presented architecture (list of banks) still helps to filter only relevant articles and comments from external sources, which helps to bring only data needed in further analysis and visualizations.

According to Eler et al. (2018), pre-processing steps through various methods have strong impact on text-mining techniques. Through lowercasing the whole text, all tokens are converted into lowercase, where some mistakes can happen (e.g. converting of abbreviation "US" into the pronoun "us"). To reduce the noise in the text, there are various techniques like deletion of double spaces, numbers, names (if needed), punctuation, rare words, stop-words and so forth. The following step to reduce dimensionality is to introduce stemming or lemmatization tasks on keywords in order to gather all variations of specific keywords (example: "bank", "banking", "banks" → "bank"). Lemmatization uses PoS (Part-of-Speech) tagging to identify grammatical categories. This feature can be useful in the parsing algorithms to detect the correct POS word or to extract the sequence of words (n-grams). Many text-mining tools use stemming which uses cutting of affixes: "banking" → "bank" + "ing".

This feature is useful later on, especially for counting or online presence metric. This metric practically counts how many mentions of a specific keyword there are for the specific page name or username. This online presence metric can be used for institution comparison in financial sector (bank1 vs. bank2). We can see trends in time (bank1 has higher online presence in last month than bank2 and by how much).

Figure 6.1 depicts the presence of targeted keywords (name of the competing banks) on all sources (Facebook, Twitter and websites) in one month. Bank 1, Bank 7 and Bank 12 have the majority of total online presence. A useful technique is to extend dictionaries by related terms and thus include other or similar concepts of the same word.

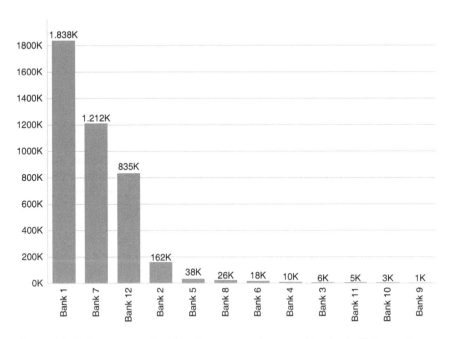

Figure 6.1 Online presence for 12 banks on targeted sources (Facebook, Twitter, relevant financial websites), during one month

Source: Authors' work.

Named entity recognition

Named entity recognition represents an important step in text mining (Saju and Shaja, 2017) and is used on large corpora, which can be used in information retrieval, information extraction and further in natural language processing, machine translation, question answering system, speech recognition, natural language generation, chatbot conversation, machine learning, document indexing, image recognition and so forth. Many industries use named entity recognition on big datasets. Most of named entity recognition techniques are based on supervised learning, which requires a large dataset in order to train a good classifier.

Named entity recognition is a process which labels a Name, that is, a sequence of words in documents, which can denote email, amounts of currency, a company/bank/institution name, brand name, city and state name, time or others (Grishman and Sundheim, 1996). The three universally accepted name entities are person, location and organization. Named entity recognition consists of mainly of two steps – detection of names in the text and classification by the type of entity – but also of discovering relationships among entities. In the detection process, problems of segmentation can appear (e.g. the "National Bank of Croatia" which is a single name, instead of "Croatia", which is a location). Named entity recognition could have its business value in industrial applications, in bank transaction details, to detect contracts, email, machine translation, question answering, spell checking and so forth. Alvarado et al. (2015) conducted named entity recognition analysis on financial documentation and publicly available non-financial dataset to extract information of risk assessment.

Financial institutions also use named entity recognition for internal data, such as extraction of client names, bank account numbers and IBAN (International Bank Account) number. An example of named entity recognition on external data (websites) is presented below:

> Shares in the green include Company1 [ORGANIZATION] as this retailer beat on earnings on an adjusted basis, with revenue topping estimates as well, Company2 [ORGANIZATION] as the Chicago-based company beat on earnings by a healthy margin, although it missed slightly on revenue, and Company3 [ORGANIZATION] – shares soaring on the cloud company's public debut, after initialling pricing at $54 [NUMBER] a share in October of 2017 [TIME]. A good day for the CEO John Smith [PERSON].

There are dictionaries with pre-defined named entities that every organization can use for quick start and result. For better results, solutions that are more complex are needed. Since tweets and comments from social media and websites usually lack context and are noisy, there are more complex solutions like supervised approach for named entity recognition (Ritter et al., 2011).

Gender prediction

Information about gender is often useful, especially when the emphasis of analysis is marketing planning and/or better understanding of customers. The simple approach in solving this problem is to make a dictionary of female and male names and then match that dictionary with user names. This can be a good approach if fast results are needed. Still, when social media and websites are analyzed, the problem arises related to the number of accounts from different organizations, bots and fake accounts with random names. In that case, the presented approach will only recognize what is in the dictionary, thus lowering the probability of recognizing the gender of the customer.

To solve this limitation, the next step would be to use natural language processing models such as "bag of words", n-grams or a combination of both. This approach analyses word usage and differences between them and differences between styles. A disadvantage again occurs with data extracted from social media. Features used for this classification task are words (authors suggest that binary representation is more effective – the word either exists or it does not in a document), average word or sentence length, POS tags (noun, verb, adjective and adverb) and word factor analysis – finding groups of similar words (e.g. there are 20 lists – example of conversation list is "known", "care", "friend" "saying") (Zhang and Zhang, 2010). Information gain is used as feature selection and with SVM (Support Vector Machine) as classifier, with the accuracy above 72%. The latest approach is usage of deep learning techniques to cope with this problem, with accuracy of over 85% (Bartle and Zheng, 2015). Lotto (2018) used various determinants, gender prediction among them, to predict financial inclusion and compared it with traditional banking services.

Sentiment analysis

Sentiment analysis or opinion analysis is used in the financial sector to identify the voice of customers. Sentiment analysis (Pang and Lee, 2008) refers to text analysis or natural language processing techniques, which helps the determination of a writer's attitude towards a specific topic. Usage of sentiment analysis is frequent in the financial domain. Nopp and Hanbury (2015) used sentiment analysis to detect risks in the banking system. Srivastava and Gopalkrishnan (2015) analyzed sentiments for the banking sector in order to assess functioning of the bank. Nyman et al. (2018) use the emotional finance hypothesis, according to which individuals take positions in financial markets and create narratives which express excitement, anxiety or losses. These narratives are created and disseminated in social interaction.

There are several approaches to building an accurate sentiment model. Some approaches address this problem from the natural language processing view, others from the machine-learning view or, recently, more specifically as a deep learning problem.

The first approach, based on natural language processing, is to build a dictionary of known negative and positive words. For this task, only extreme polarities and words that can be correctly associated with the polarity. Based on the developed dictionary, the sentiment is calculated by a simple count of words found in a specific document from the dictionary. The polarity with more discovered words "wins", and the text is then classified.

The next approach, based on machine learning, would be to create a large dataset containing documents that are first classified manually (by humans). Based on the classification, a machine-learning model can be developed that can provide rules for automated classification. Problem can be addressed as classification of two classes (positive or negative) or more (e.g. range from 1 to 5 for sentiment intensity). Features can be unigrams, bigrams, or combination of both (Go et al., 2009). A document term matrix is built based on our features and values in this matrix, which can be either frequencies like TF (term frequency), TF-IDF (term frequency–inverse document frequency) or binary representation. In our example of big data architectures, a machine-learning model can be used on batch data but also on real-time data in order to perform real-time classification. Accuracy can be greater than 80% even with simple algorithms with correct feature selection and a noise removal process (Narayanan et al., 2013).

The last approach, based on deep learning, the sentiment analysis would be performed using word embeddings, such as word2vec or GloVe (Zhang et al., 2018). Word embeddings are used to represent words as vectors. With this technique, similar words can be mapped to nearby points in continuous vector space. Deep learning is an improvement of other approaches and especially in sentiment classification of relatively small documents (tweets, comments) as in our case. Deep learning is even used for tasks that are hard to solve, like irony or sarcasm detection (Mandelbaum and Shalev, 2016).

Figure 6.2 presents the sentiment analysis over time for one bank. Sentiment values are from 0 to 10 where 0 is bad, 10 is good and 5 would be neutral. We can see that sentiment trend (dotted line) for this bank is rising in the selected period (October 2017 to February 2018). Sentiment analysis can be combined with other text-mining methods, and Figure 6.3 presents one such example combining sentiment analysis for one bank by gender.

Topic extraction

Topic modelling or topic prediction/extraction is based on the number and distribution of terms across documents by counting the probability of belonging to the certain topic. Moro et al. (2015) performed topic detection of a large number of manuscripts using text-mining techniques when detecting terms belonging to business intelligence and banking domains. They used a latent Dirichlet allocation (LDA) model to detect topics by using a dictionary of terms in order to detect topics and research directions. They grouped articles into several relevant topics, followed by dictionary analysis to identify relations between terms and topics of grouping articles. This research showed that credit banking was the main trend with topics of risk, fraud detection, credit approval and bankruptcy. For each document, it is possible to obtain the probability of belonging to the certain topic. In this way, it is possible to identify topics capturing more attention.

Data from social media can be used to find discussed topics in a certain time period. Previous research indicates that these data can be a good source of entity-oriented topics that have low coverage in traditional media news (e.g. Zhao et al.,

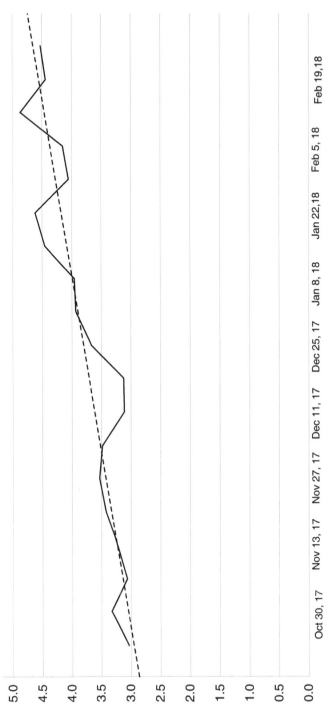

Figure 6.2 Sentiment analysis over time for one bank

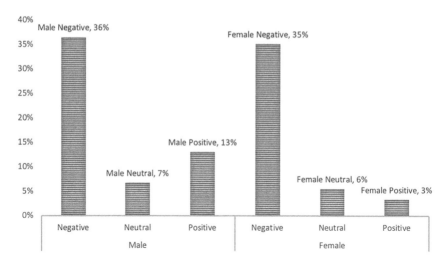

Figure 6.3 Sentiment analysis for one bank by gender

2011). Input in the model should be a matrix of document-terms format with TF-IDF frequencies as values or binary representation (0 or 1). A common approach for topic extraction is the unsupervised machine-learning approach. Like in previous examples, each document that is ingested into our hypothetical system can be assigned to a specific topic. Usually topics are time-related. Topics are usually generated for a specific time, like one week, one month or one year.

New approaches take also deep learning techniques for topic extraction. Popular word embedding in this case is lda2vec, which is a modification of word2vec presented in sentiment analysis (Moody, 2016). Lda2vec uses word2vec principles, expanding it to word, document and topic vectors. Topic extraction helps in answering the question "WHAT" is discussed about our institution or our competitors. Topics are usually represented as word clouds, but they can be visualized by some more complex graphical representation (LDAvis, where an intertopic distance map is visualized with principal component analysis (PCA)). Won Sang and So Young (2017) proposed a framework to identify the rise and fall of emerging topics in the financial industry using abstracts of financial business model patents in order to discover topics from documents, aiming to enable understanding of the changing trends of financial business models over time.

Social network analysis

Social network analysis is a different type of analysis from text analysis, but it is used here to show how text analysis and its result can be integrated with it (L'Huillier et al., 2011). With this approach, we can enrich our previous analysis like topic extraction. One way is to combine the SNA graph with topic extraction in order to find out which user in our specific network belongs to the topic generated in the previous section.

Social network analysis (SNA) is the process that is based on graph theory and used for better understanding of social structures. Since in our hypothetical case we collect data from social media and the web, we can use data to describe interactions between users from those sources (e.g. Twitter friends and followers). When we talk about SNA, we talk about nodes and edges. In our case (e.g. Twitter), each node would be one Twitter user, and each edge is the relationship between two users (user is connected to other user by follow or retweet). Usual metrics calculated with SNA techniques are centrality measures, node degrees (used to find users who are highly connected), closeness (goal is to find users who can spread information to others), clustering coefficient and PageRank (Ediger et al., 2010).

The next example of usage is when we identify a user who can spread a message easily on our network of interest (with SNA techniques). Then we can use textual data from followers of that user to find out common interests. This information can be used for marketing campaigns to generate best keywords. CGI Group (2011) used SNA to perform fraud detection, and Morales et al. (2014) estimated bank financial strength during the financial crisis.

Conclusion

Text mining can be integrated into business intelligence process and business applications, which represents a promising target. Many companies and institutions have a large amount of data, which can be exploited and used to extract data and create information linked into new knowledge. In this research process, text mining techniques can help in customer analytics, marketing opportunities, fraud prevention, improving operational activities and developing new business models. Fan et al. (2006) suggest future development in integration of data and text mining used to discover hidden information in different types of documents collected from various resources, with the purpose of improved decision-making.

Big data text mining is used by the financial sector in many fields to improve customer relationships, marketing campaigns and customer segmentation. This paper analyzed big data architecture and text mining techniques for the financial sector; the outcome is presented below. First, customer-related data is ingested from popular social media (Facebook and Twitter) and web (popular web pages) with big data technologies. This is used as a foundation for text analysis. Second, we used text mining to extract valuable insights from our collected datasets. Third, we presented the most often used text-mining methods, and most of them can be implemented easily in any financial institution. There are numerous approaches and techniques, and finding the right one depends on the needs of the financial institution.

This study can be further extended by covering presented techniques with a real dataset from a specific country. In addition, instead of concentrating only on text analysis, other case studies and techniques can be presented like fraud detection, churn analysis, risk analysis, security intelligence and so forth.

References

Alexander, L., Das, S. R., Ives, Z., Jagadish, H. V., and Monteleoni, C. (2017). Research Challenges in Financial Data Modeling and Analysis. *Big Data*, 5(3), 177–188.

Alvarado, J.C.S., Verspoor, K., and Baldwing, T. (2015). Domain Adaptation of Named Entity Recognition to Support Credit Risk Assessment. *Proceedings of Australasian Language Technology Workshop*, 84–90.

Banerjee, N. (2016). Artificial Intelligence – Transforming the Financial Services Industry. *Niveshak*, 9(11), 19–22.

Bartle, A., and Zheng, J. (2015). Gender Classification with Deep Learning. Available at: https://cs224d.stanford.edu/reports/BartleAric.pdf / 12th August 2018.

Bharti, S. K., and Babu, K. S. (2017). Automatic Keyword Extraction for Text Summarization: A Survey. Available at: https://arxiv.org/ftp/arxiv/papers/1704/1704.03242.pdf / 12th August 2018.

Bholat, D., Hansen, S., and Pedro, S. (2015). Text Mining for Central Banks: Handbook. *Centre for Central Banking Studies*, (33), pp. 1–19.

CGI Group Inc. (2011). Implementing Social Network Analysis for Fraud Prevention. Available at: www.cgi.com/sites/default/files/white-papers/Implementing-social-network-analysis-for-fraud-prevention.pdf / 12th August 2018.

Davison, H., Watkins, T., and Wright, M. (1989). Developing New Personal Financial Products – Some Evidence of the Role of Market Research. *International Journal of Bank Marketing*, 7(1), 8–15.

Dunđer, I., Seljan, S., and Stančić, H. (2015). Koncept automatske klasifikacije registraturnoga i arhivskoga gradiva. *Zaštita arhivskoga gradiva u nastajanju*, pp. 195–211.

Ediger, D., Jiang, K., Riedy, J., Bader, D. A., and Corley, C. (2010, September). Massive Social Network Analysis: Mining Twitter for Social Good. In *39th International Conference on Parallel Processing*, San Diego, CA, 2010, pp. 583–593.

Eler, D. M., Grosa, D., Pola, I., Garcia, R., Correia, R., and Teixeira, J. (2018). Analysis of Document Pre-Processing Effects in Text and Opinion Mining. *Information*, 9(4), p. 100.

Fan, W., Wallace, L., Rich, S., and Zhang, Z. (2006). Tapping the Power of Text Mining. *Communications of the ACM*, 49(9), pp. 77–82.

Finacle Connect (2018). Connecting the Banking World. Artificial Intelligence Powered Banking. Available at: https://active.ai/wp-content/uploads/2018/05/Finacle-Connect-2018-leading-ai-online.pdf / 12th August 2018.

Financial Stability Board. (2017). Artificial Intelligence and Machine Learning in Financial Services: Market Developments and Financial Stability Implications. Available at: www.fsb.org/wp-content/uploads/P011117.pdf/12th August 2018.

Furht, B., and Villanustre, F. (2016). Introduction to Big Data. In *Big Data Technologies and Applications*, Ed. B. Furth and F. Villanustre. Springer, Cham, pp. 3–11.

Gandomi, A., and Haider, M. (2015). Beyond the Hype: Big Data Concepts, Methods, and Analytics. *International Journal of Information Management*, 35(2), pp. 137–144.

Gašpar, A., and Seljan, S. (in print). Consistency of Translated Terminology Measured by the Herfindahl-Hirshman Index (HHI). *Lecture Notes in Computer Science (LNCS)*.

Go, A., Bhayani, R., and Huang, L. (2009). Twitter Sentiment Classification using Distant Supervision. CS224N Project Report, Stanford, 1(12). Available at: www.yuefly.com/Public/Files/2017-03-07/58beb0822faef.pdf / 12th August 2018.

Grishman, R., and Sundheim, B. (1996). Message Understanding Conference-6: A Brief History. In *COLING 1996 Volume 1: The 16th International Conference on Computational Linguistics*, pp. 466–471.

Hasan, K. S., and Ng, V. (2014). Automatic Keyphrase Extraction: A Survey of the State of the Art. In *Proceedings of the 52nd Annual Meeting of the Association for Computational Linguistics*, 1, pp. 1262–1273.

He, W., Zha, S., and Li, L. (2013). Social Media Competitive Analysis and Text Mining: A Case Study in the Pizza Industry. *International Journal of Information Management*, 33(3), pp. 464–472.

Kloptchenko, A., Eklund, T., Karlsson, J., Back, B., Vanharanta, H., and Visa, A. (2004). Combining Data and Text Mining Techniques for Analysing Financial Reports. *Intelligent Systems in Accounting, Finance and Management*, 12(1), pp. 29–41.

Kreps, J. (2014). Questioning the Lambda Architecture. Available at: www.oreilly.com/ideas/questioning-the-lambda-architecture.

L'Huillier, G., Alvarez, H., Ríos, S.A., and Aguilera, F. (2011). Topic-based Social Network Analysis for Virtual Communities of Interests in the Dark Web. *ACM SIGKDD Explorations Newsletter*, 12(2), pp. 66–73.

Lotto, J. (2018). Examination of the Status of Financial Inclusion and its Determinants in Tanzania. *Sustainability*, 10(8), p. 2873.

Mak, K.Y.M., Ho, G.T.S., and Ting, S.L. (2011). A Financial Data Mining Model for Extracting Customer Behavior. *International Journal of Engineering Business Management*, 3(3), pp. 59–72.

Mandelbaum, A., and Shalev, A. (2016). Word Embeddings and their Use in Sentence Classification Tasks. Available at: https://arxiv.org/abs/1610.08229.

Marz, N., and Warren, J. (2015). *Big Data: Principles and Best Practices of Scalable Real-time Data Systems*. Manning Publications, New York.

Mathew, S. (2012). Financial Services Data Management: Big Data Technologies in Financial Services. Oracle White Paper. Available at: www.oracle.com/us/industries/financial-services/bigdata-in-fs-final-wp-1664665.pdf / 12th August 2018.

Moody, C.E. (2016). Mixing Dirichlet Topic Models and Word Embeddings to Make lda2vec. Available at: https://arxiv.org/abs/1605.02019 / 12th August 2018.

Morales, M., Brizan, D., Ghaly, H., Hauner, T., Ma, M., Reza, S., and Rosenberg, A. (2014). Application of Social Network Analysis in the Estimation of Bank Financial Strength during the Financial Crisis. NLP Unshared Task in PoliInformatics. Available at: http://thomashauner.com/papers/NLPunsharedTaks_application%20social%20network%20(2014).pdf/ 12th August 2018.

Moro, S., Cortez, P., and Rita, P. (2015). Business Intelligence in Banking: A Literature Analysis from 2002 to 2013 using Text Mining and Latent Dirichlet Allocation. *Expert Systems with Applications*, 42(3), pp. 1314–1324.

Narayanan, V., Arora, I., and Bhatia, A. (2013). Fast and Accurate Sentiment Classification Using an Enhanced Naive Bayes Model. In *International Conference on Intelligent Data Engineering and Automated Learning*, Springer, Berlin, Heidelberg, pp. 194–201.

Nopp, C., and Hanbury, A. (2015). Detecting Risks in the Banking System by Sentiment Analysis. In *Proceedings of the 2015 Conference on Empirical Methods in Natural Language Processing*, pp. 591–600.

Nyman, R., Kapadia, S., Tuckett, D., Gregory, D., Ormerod, P., and Smith, R. (2018). News and Narratives in Financial Systems: Exploiting Big Data for Systemic Risk Assessment. Available at: www.bankofengland.co.uk/working-paper/2018/news-and-narratives-in-financial-systems / 12th August, 2018.

Pang, B., and Lee, L. (2008). Opinion Mining and Sentiment Analysis. *Foundations and Trends® in Information Retrieval*, 2(1–2), pp. 1–135.

Ritter, A., Clark, S., and Etzioni, O. (2011). Named Entity Recognition in Tweets: An Experimental Study. In *Proceedings of the Conference on Empirical Methods in Natural Language Processing, Association for Computational Linguistics*, pp. 1524–1534.

Saju, J.C., and Shaja, A.S. (2017). A Survey on Efficient Extraction of Named Entities from New Domains Using Big Data Analytics. In *2nd International Conference on Recent Trends and Challenges in Computational Models (ICRTCCM)*, pp. 170–175.

Seljan, S., Dunđer, I., and Gašpar, A. (2013). From Digitisation Process to Terminological Digital Resources. In *Information & Communication Technology Electronics & Microelectronics (MIPRO) Proceedings*, pp. 1053–1058.

Seljan, S., Baretić, M., and Kučiš, V. (2014). Information Retrieval and Terminology Extraction in Online Resources for Patients with Diabetes. *Collegium Antropologicum*, 38(2), pp. 705–710.

Seljan, S., Dalbelo Bašić, B., Šnajder, J., Delač, D., Šamec-Gjurin, M., and Crnec, D. (2009b). Comparative Analysis of Automatic Term and Collocation Extraction. In *INFuture 2009 Conference Proceedings, Digital resources and Knowledge*, pp. 219–228.

Seljan, S., and Gašpar, A. (2009a) First Steps in Term and Collocation Extraction from English-Croatian Corpus. In *Proceedings of 8th International Conference on Terminology and Artificial Intelligence*. Toulouse, France. Available at: http://ceur-ws.org/Vol-578/paper21.pdf / 12th August 2018.

Seljan, S., Stančić, H., and Dunđer, I. (2017). Extracting Terminology by Language Independent Methods. In *Forum Translationswissenschaft. Translation Studies and Translation Practice*. Peter Lang GmbH, Frankfurt am Main, pp. 141–147.

Srivastava, U., and Gopalkrishnan, S. (2015). Impact of Big Data Analytics on Banking Sector: Learning for Indian Banks. *Procedia Computer Science*, 50, 643–652.

Turner, D., Schroeck, M., and Shockley, R. (2012). *Analytics: The Real-world Use of Big Data in Financial Services*. IBM Institute for Business Value in Collaboration with Said Business School, University of Oxford.

Won Sang, L., and So Young, S. (2017). Identifying Emerging Trends of Financial Business Method Patents. *Sustainability*, 9(9), 1670.

Zhang, C., and Zhang, P. (2010). *Predicting Gender from Blog Posts*. Amherst, USA: University of Massachusetts.

Zhang, L., Wang, S., and Liu, B. (2018). Deep Learning for Sentiment Analysis: A Survey. Available at: https://arxiv.org/abs/1801.07883 / 12th August 2018.

Zhao, D. (2013). Frontiers of Big Data Business Analytics: Patterns and Cases in Online Marketing. In *Big Data and Business Analytics*, Ed. J. Leibowitz. CRC Press, Boca Raton, pp. 46–68.

Zhao, W.X., Jiang, J., Weng, J., He, J., Lim, E.P., Yan, H., and Li, X. (2011, April) Comparing Twitter and Traditional Media Using Topic Models. In *European Conference on Information Retrieval*. Springer, Berlin, Heidelberg, pp. 338–349.

7 CEL

Citizen economic level using SAW

Andino Maseleno, K. Shankar, Miftachul Huda, Marini Othman, Prayugo Khoir and Muhammad Muslihudin

Introduction

Indonesia is the country with the largest population, and the most populous after China, but in terms of wealth and income, Indonesia is still inferior. Wealth is something that cannot be calculated only by visits from the outside. It is important for an owner of wealth to know how much is owned in order to better control revenue and expenditure. This opportunity becomes the outstanding value to give insights into the demands with the trends of the marketplace [1]–[4]. The Trade of Act. Chapter 1 General Provisions. Article 3. includes about grain-trading activities aimed at setting bouts, improving national economic growth, increasing the use and trade of domestic products, increasing employment and create jobs, and ensuring smooth distribution and availability of staples and essential items.

The division of economic levels in Indonesia are weak economy, medium, middle down, and middleclass that further efforts are doing useful measures to control property of owned because just to calculate it manually is not enough. This would be the potential to enhance market innovation in the digital age [5]–[8]. The existing results will not be accurate, and there will only be mistakes. Therefore an application is needed that can calculate the property and assets owned by the classification level of the economy [9]–[12].

The Indonesian government always provides assistance to people with different needs. Types of assistance include the provision of rice for the poor, BPJS (Badan Penyelenggara Jaminan Sosial, or the Social Insurance Administration Organization) and so forth. Errors that occur can be reduced, especially if there are no mistakes on the goals that should be addressed properly, but because of mistakes and ignorance, the targets that should be intended [13]–[16]. The application is made on the goals already in mind, which are to help facilitate the distribution of Raskin, specify recipients of health cards like BPJS, improve accuracy or timeliness based on the data already acquired, calculate the amount of wealth, and make it easier for people to determine the amount of tax. The point of managing wealth initially has to be engaged with an appropriate manner and a wise approach [17]–[20], in accordance with the Law of Republic of Indonesia Number 36 Year 2008 on the Fourth Amendment of Law Number 7 of 1983 Income Tax, contained in Article 2. Another purpose of this research was to classify economic classes based on an economic level that has been determined. Based on the explanation above, formulated problems can be solved to calculate the economy by classifying it into several levels based on a few criteria/variables that have been set.

Decision support system

The decision support system (DSS) is a system to support decision makers in solving semi-structured problems. DSS aims as a tool for decision makers to expand their abilities, but not to replace their judgments [21]–[24]. Man and Watson propose that a DSS is an interactive system that helps decision makers through the use of semi-structured and unstructured data and decision models [25]–[28]. It proposes that the types of decisions are grouped into two kinds. The first is the programmed decision, this decision is related to a known issue, and the decision making is routine and scheduled [29]–[33]. The moment the decision is not programmed, the decision is new because it deals with new problems [34]–[37].

There are three components of a decision support system. The first is that data management serves as a provider of data required by the system [38]–[39]. The second management model, the base management model, interacts well with the user interface to obtain commands from data management to get the data to be processed [40]–[43]. The third user interface is the most important component in the decision support system because it interacts between users of the system, either to enter information in the system or display information to the user [44]–[47].

Uncertainty reasoning

Uncertainty is a concept that reflects human lack of sureness about something or someone. Reasoning theories are divided into certainty reasoning theories and uncertainty reasoning theories. Certainty thinking still prevails in different disciplines. In Cartesian philosophy, mathematics was the only accurate knowledge. With the combination of mathematics and physics, all sorts of natural and social phenomena could be explained in science [48]–[50]. Leibniz, philosopher and mathematician, was convinced that the symbolic language of science could construct the universal logic and logical calculus, and all phenomena could be clearer. With absolute time-space, Newton was sure that all the observable physical quantities in principle could be accurately measured and its foundation was the uncertainty of physical laws. An unknown world was deterministic for perfectly rational policymakers in the traditional design science view [51]–[54]. One could achieve the effect of maximization as long as marginal benefit equaled marginal cost decision. However, the world is uncertain. The lack of certainty is ubiquitous throughout the world. Uncertainty is distinguished from certainty in the degree of belief or confidence. If certainty is referred to as a perception or belief that a certain system or phenomenon can experience, uncertainty indicates a lack of confidence or trust in an article of knowledge or decision [55]–[58]. Uncertainty is a term used in subtly different ways in a number of fields, including philosophy, physics, statistics, economics, finance, insurance, psychology, sociology, engineering and information science. It applies to predictions of future events, to physical measurements that are already made or to the unknown. Uncertainty arises in partially observable and/or stochastic environments, as well as due to ignorance and/or indolence. According to the Cambridge Dictionary, uncertainty is a situation in which something is not known, or a thing that is not known or certain.

Uncertainty arises from different sources in various forms, and is classified in different ways by different communities. It is categorized into aleatory uncertainty or epistemic uncertainty. Aleatory uncertainty derives from the natural variability of the physical world and reflects the inherent randomness in nature [59]–[62]. It exists naturally regardless of human knowledge. For example, in an event of flipping a coin, the coin comes up heads or tails with some randomness. Even if researchers do many experiments and know the probability of coming up heads, researchers still cannot predict the exact result in the next turn. Aleatory uncertainty cannot be eliminated or reduced by collecting more knowledge or information. No matter whether people know it or not, this uncertainty stays there all the time. Aleatory uncertainty is sometimes also referred to as natural variability, objective uncertainty, external uncertainty, random uncertainty, stochastic uncertainty, inherent uncertainty, irreducible uncertainty, fundamental uncertainty, real-world uncertainty or primary uncertainty. Epistemic uncertainty originates from humanity's lack of knowledge of the physical world and lack of the ability to measure and model the physical world. Unlike aleatory uncertainty, given more knowledge of the problem and proper methods epistemic uncertainty can be reduced and sometimes can even be eliminated. Epistemic uncertainty is sometimes also called knowledge uncertainty, subjective uncertainty, internal uncertainty, incompleteness, functional uncertainty, informative uncertainty or secondary uncertainty. The Dempster-Shafer mathematical theory of evidence can deal with both aleatory and epistemic uncertainty.

Fuzzy logic

Fuzzy logic can handle problems with imprecise data and give more accurate results. Professor L.A. Zadeh introduced the concept of fuzzy logic; soon after, researchers used this theory for developing new algorithms and decision analysis. Fuzzy logic has been applied successfully in hundreds of application domains including image processing, mobile robot navigation, distance relation, high energy physics, natural numbers, medicinal chemistry, robot manipulators, optimization [63]–[77] of machining processes, power converters, control of permanent-magnet synchronous motors and electric power systems. Fuzzy sets, proposed by Zadeh [36] as a framework to encounter uncertainty, vagueness and partial truth, represents a degree of membership for each member of the universe of discourse to a subset of it. The original motivation for fuzzy logic is to provide the basis for reasoning under non-binary information. The ensuing reasoning system this is often referred to is approximate reasoning or fuzzy reasoning. However, this should not be taken to imply that the resulting system is any less exact than that afforded by crisp logic. Indeed, fuzzy reasoning might be considered more exact precisely because it does not assume a binary universe. The basis for formal reasoning is an inference procedure, itself based upon an appropriate model for "if-then" rules, or modus ponens. The general goal is to infer the degree of truth associated with a proposition, B, from the implication, A, or $A \mathbin{!} B$. Consider, "A" denotes "sharp corner" and "B" denotes "approach slowly", than the implication can naturally express by, premise 1 (fact): x is A; premise 2 (fact): IF x is A THEN y is B; consequence (conclusion): y is B or premise 1 (fact): x is A'; premise 2 (fact): IF x is A THEN y is B; consequence (conclusion): y is B'.

On the basis of the description of input and output variables, this research has constructed 25 rules, which is a series of if-then statements.

Now, in a position to build some reasoning engines, consider the following three special cases:

1   Single rule with single premise. The premise simplifies to the special case of a scalar thresholds, or $\mu B0 = w^\wedge \mu B(y)$.
2   Single rule with multiple premises. The premise simplifies to the special case of the minimum of two scalar thresholds.
3   Multiple rules with multiple premises. At this point, the basis for fuzzy reasoning with the remaining problem of establishing what the fuzzy consequent actually means in practice.

Fuzzy multiple attribute decision making

Fuzzy Multiple Attribute Decision Making (FMADM) is a method to determine the weight value in each attribute, and then proceed into the ranking process that will select the alternatives already given. Basically, there are three kinds of approaches for finding attribute weights: subjective approach, objective approach and integration approach. Between subjective and objective, each has its own way to determine the weight value of each alternative. In a subjective approach, the weighted value is determined by the subjectivity of the decision makers, so that several factors in the alternative ranking process can be freely determined. As for the objective approach, the weighted value is mathematically calculated so that it ignores the subjectivity of the decision maker. There are several methods that can be used for FMADM completion, among others:

1   Simple Additive Weighting (SAW)
2   Weighted Product (WP)
3   ELECTRE
4   Technique for Order Preference by Similarity to Ideal Solution (TOPSIS)
5   Analytic Hierarchy Process (AHP).

Simple additive weighting method

The simple additive weighting (SAW) method is also known as the term weighted summation method. Fuzzy logic can handle problems with imprecise data and give more accurate results in terms of providing opportunities in a way to maximize the potential data value collected within the big data approach. SAW is related to uncertainty. Some research has been done related to uncertainty and sport and uncertainty related with other fields. The basic concept of the SAW method is to find the sum of the weighted performance rating for each alternative on all attributes. The SAW method requires a process of normalizing the decision matrix (X) to a scale that can be compared with all the ratings of existing alternatives.

$$r_{ij} = \begin{cases} \dfrac{x_{ij}}{\underset{i}{\text{Max }} x_{ij}} \\[4ex] \dfrac{\underset{i}{\text{Min }} x_{ij}}{x_{ij}} \end{cases} \tag{7.1}$$

where:

r_{ij} = value normalized performance rating
$\text{Max } x_{ij}$ = the maximum value of each row and column
$\text{Min } x_{ij}$ = minimum value of each row and column
X_{ij} = rows and columns of a matrix
With r_{ij} being the normalized performance rating of alternatives on attribute Ai Cj; $i = 1,2, \ldots$ m and $j = 1,2, \ldots,$ n.

The preference value for each alternative (Vi) is given as:

$$V_i = \sum_{J=1}^{n} W_j \, r_{ij} \tag{7.2}$$

V_i = value preferences
W_j = weight rankings
r_{ij} = normalized performance rating
V_i = larger value indicates that the alternative Ai is selected.

Step Completion Simple Additive Weighting (SAW):

1 Determine the criteria that will be used as a reference in the decision, namely Ci.
2 Determine the suitability rating each alternative on each criterion.
3 Make a decision matrix based on criteria (Ci), then normalized matrix based on the equations adjusted for the type attribute (attribute or attributes benefit costs) that is required to be normalized matrix R.
4 The final result is obtained from the process of ranking the summation of the matrix multiplication R normalized with the weight vector in order to obtain the greatest value is selected as the best alternative (Ai) as a solution.

Selection criteria analysis

In this research, the process of selecting candidates, the receiver uses the Multi-Attribute Fuzzy Decision Making method. There are five criteria used: C1 = the number of people living in one house; C2 = amount of the last vehicle tax; C3 = total last electricity bill; C4 = total monthly income; C5 = total last property tax.

These criteria have proven authenticity and a high level of security that is difficult to forge. These criteria will then be used as input in the steps of the selection process by the method FMADM. From each of these criteria will be determined, weight-weight. On the weight of the five fuzzy numbers, namely very low, low, simply, high and very high.

Implementation

At this stage, the researchers took existing criteria to be used as an indicator in choosing academic officials and the criteria will be translated into fuzzy numbers whose values are the criteria to be given some weight. The weight of the five fuzzy numbers are (0) very low, (0.25) low, (0.50) fair, (0.75) high and (1) very high. Table 7.1 shows Criteria 1 (C1) the number of people staying, Table 7.2 shows Criteria 2 (C2) total tax charge last vehicles, Table 7.3 shows Criteria 3 (C3) total cost of electricity bills, Table 7.4 shows Criteria 4 (C4) monthly incomes, and Table 7.5 shows Criteria 5 (C5) tax on land and buildings.

Table 7.1 Criteria 1 (C1): the number of people staying

Number of People (C1)	Crisp Numbers
$C1 \geq 1 \leq 5$	0
$C1 \geq 6 \leq 10$	0.25
$C1 \geq 11 \leq 20$	0.50
$C1 \geq 21 \leq 50$	0.75
$C1 > 100$	1

Table 7.2 Criteria 2 (C2): total tax charge last vehicles

Vehicle Tax (C2)	Crisp Numbers
$C2 \geq 0$ IDR	0
$C2 \leq 100,000$ IDR $\leq 500,000$ IDR	0.25
$C2 \geq 501,000$ IDR $\leq 1,000.000$ IDR	0.50
$C2 \geq 1,001,000$ IDR $\leq 10,000,000$ IDR	0.75
$C2 > 10,001,000$ IDR	1

Table 7.3 Criteria 3 (C3): total cost of electricity bills

Electricity Accounts (C3)	Crisp Numbers
C3 ≥ 0 IDR	0
C3 ≤ 1,000 IDR ≤ 50,000 IDR	0.25
C3 ≥ 51,000 IDR ≤ 500,000 IDR	0.50
C3 ≥ 501,000 IDR ≤ 1,000,000 IDR	0.75
C3 > 1,001,000 IDR	1

Table 7.4 Criteria 4 (C4): monthly incomes

Monthly Incomes (C4)	Crisp Numbers
C4 ≥ 0 IDR	0
C4 ≤ 100,000 IDR ≤ 500,0000 IDR	0.25
C4 ≥ 501,000 IDR ≤ 1,000,000 IDR	0.50
C4 ≥ 1,001,000 IDR ≤ 10,000,000 IDR	0.75
C4 > 10,001,000 IDR	1

Table 7.5 Criteria 5 (C5): tax on land and buildings

Tax on Land and Buildings (C5)	Crisp Numbers
C5 ≥ 0 IDR	0
C5 ≤ 50,000 IDR ≤ 100,000 IDR	0.25
C5 ≥ 101,000 IDR ≤ 500,000 IDR	0.50
C5 ≥ 501,000 IDR ≤ 1,000,000 IDR	0.75
C5 >1,001,000 IDR	1

Results and discussion

At the start screen, a form appears containing the data to be entered; delete buttons are provided for users. Figure 7.1 shows program application of class economic. Input the calculation in the form of charging the number of people staying, vehicle tax, last electricity bill, income per month and last property taxes. Then click PROCESS, and the results will be shown.

Figure 7.2 shows a printout of the form in print preview of a class program economic, used as evidence or a memorandum.

Conclusion

Based on the results of the previous discussion, it could be concluded that the resulting decision support system can be used to determine the level of the economy by the classification of the results already obtained. From the discussion based on those alternatives, the following classifications were obtained: V1

Figure 7.1 Program application of level of economic

(bottom or poor) with the value points 0:27, V2 (medium up or medium) with the value points 0.93, V3 (rich) with the value points 1:45. The system is built to help speed up the process for the user, to make it more effective and efficient. Future development of this work will be better using the machine learning method.

References

[1] Adela, A., Jasmi, K.A., Basiron, B., Huda, M. and Maseleno, A. (2018). Selection of Dancer Member using Simple Additive Weighting. *International Journal of Engineering & Technology,* 7(3), 1096–1107.

[2] Aminin, S., Huda, M., Ninsiana, W. and Dacholfany, M.I. (2018). Sustaining Civic-Based Moral Values: Insights from Language Learning and Literature. *International Journal of Civil Engineering and Technology*, 9(4), 157–174.

[3] Amin, M.M., Nugratama, M.A.A., Maseleno, A., Huda, M. and Jasmi, K. A. (2018). Design of Cigarette Disposal Blower and Automatic Freshener using Mq-5 Sensor

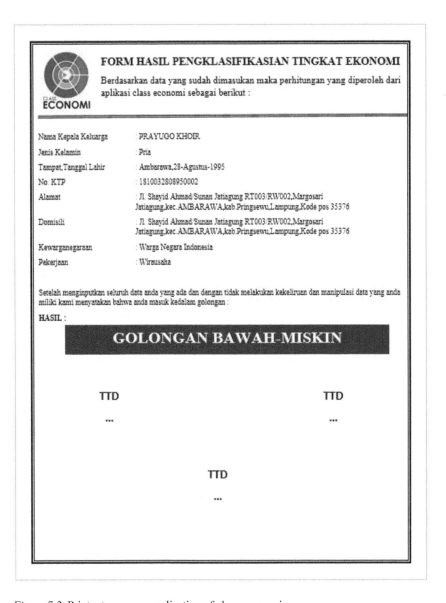

Figure 7.2 Printout program application of class economic

Based on Atmega 8535 Microcontroller. *International Journal of Engineering & Technology*, 7(3). 1108–1113.

[4] Anshari, M., Almunawar, M. N., Shahrill, M., Wicaksono, D. K. and Huda, M. (2017). Smartphones Sage in the Classrooms: Learning Aid or Interference?. *Education and Information Technologies*, 22(6), 3063–3079.

[5]   Atmotiyoso, P. and Huda, M. (2018). Investigating Factors Influencing Work Performance on Mathematics Teaching: A Case Study. *International Journal of Instruction*, 11(3), 391–402.

[6]   Anggraeni, E. Y., Huda, M., Maseleno, A., Safar, J., Jasmi, K. A., Mohamed, A. K., Hehsan, A., Basiron, B., Ihwani, S. S., Wan Embong, W. H., Mohamad, A. M., Mohd Noor, S. S., Fauzi, A. N., Wijaya, D. A. and Masrur, M. (2018). Poverty Level Grouping using SAW Method. *International Journal of Engineering and Technology*, 7(2.27), 218–224.

[7]   Huda, M. and Kartanegara, M. (2015). Islamic Spiritual Character Values of al-Zarnūjī's Ta 'līm al-Muta 'allim. *Mediterranean Journal of Social Sciences*, 6(4S2), 229–235.

[8]   Huda, M., Anshari, M., Almunawar, M. N., Shahrill, M., Tan, A., Jaidin, J. H., . . . & Masri, M. (2016a). Innovative Teaching in Higher Education: The Big Data Approach. *The Turkish Online Journal of Educational Technology*, 15(Special issue), 1210–1216.

[9]   Huda, M., Yusuf, J. B., Jasmi, K. A. and Nasir, G. A. (2016b). Understanding Comprehensive Learning Requirements in the Light of al-Zarnūjī's Ta'līm al-Muta'allim. *Sage Open*, 6(4), 1–14.

[10]   Huda, M., Yusuf, J. B., Jasmi, K. A. and Zakaria, G. N. (2016c). Al-Zarnūjī's Concept of Knowledge ('ilm). *SAGE Open*, 6(3), 1–13.

[11]   Huda, M., Jasmi, K. A., Mohamed, A. K., Wan Embong, W. H. and Safar, J. (2016d). Philosophical Investigation of Al-Zarnuji's Ta'lim al-Muta'allim: Strengthening Ethical Engagement into Teaching and Learning. *Social Science*, 11(22), 5516–5551.

[12]   Huda, M., Sabani, N., Shahrill, M., Jasmi, K. A., Basiron, B. and Mustari, M. I. (2017a). Empowering Learning Culture as Student Identity Construction in Higher Education. In A. Shahriar and G. Syed (Eds.), *Student Culture and Identity in Higher Education* (pp. 160–179). Hershey, PA: IGI Global. http://doi.org/10.4018/978-1-5225-2551-6.ch010.

[13]   Huda, M., Jasmi, K. A., Hehsan, A., Shahrill, M., Mustari, M. I., Basiron, B. and Gassama, S. K. (2017b). Empowering Children with Adaptive Technology Skills: Careful Engagement in the Digital Information Age. *International Electronic Journal of Elementary Education*, 9(3), 693–708.

[14]   Huda, M., Shahrill, M., Maseleno, A., Jasmi, K. A., Mustari, I. and Basiron, B. (2017c). Exploring Adaptive Teaching Competencies in Big Data Era. *International Journal of Emerging Technologies in Learning*, 12(3), 68–83.

[15]   Huda, M., Jasmi, K. A., Basiron, B., Mustari, M.I.B. and Sabani, A. N. (2017d). Traditional Wisdom on Sustainable Learning: An Insightful View From Al-Zarnuji's Ta 'lim al-Muta 'allim. *SAGE Open*, 7(1), 1–8.

[16]   Huda, M., Jasmi, K. A., Embong, W. H., Safar, J., Mohamad, A. M., Mohamed, A. K., Muhamad, N. H., Alas, Y. and Rahman, S. K. (2017e). Nurturing Compassion-Based Empathy: Innovative Approach in Higher Education. In M. Badea and M. Suditu (Eds.), *Violence Prevention and Safety Promotion in Higher Education Settings* (pp. 154–173). Hershey, PA: IGI Global. http://doi.org/10.4018/978-1-5225-2960-6.ch009.

[17]   Huda, M., Jasmi, K. A., Alas, Y., Qodriah, S. L., Dacholfany, M. I. and Jamsari, E. A. (2017f). Empowering Civic Responsibility: Insights From Service Learning. In S. Burton (Ed.), *Engaged Scholarship and Civic Responsibility in Higher Education* (pp. 144–165). Hershey, PA: IGI Global. http://doi.org/10.4018/978-1-5225-3649-9.ch007.

[18] Huda, M., Jasmi, K.A., Mustari, M.I., Basiron, B., Mohamed, A.K., Embong, W., . . . & Safar, J. (2017g). Innovative E-Therapy Service in Higher Education: Mobile Application Design. *International Journal of Interactive Mobile Technologies*, 11(4), 83–94.

[19] Huda, M., Jasmi, K.A., Mustari, M.I. and Basiron, B. (2017h). Understanding Divine Pedagogy in Teacher Education: Insights from Al-zarnuji's Ta'lim Al-Muta'Allim. *The Social Sciences*, 12(4), 674–679.

[20] Huda, M., Jasmi, K.A., Mustari, M.I.B. and Basiron, A.B. (2017i). Understanding of Wara' (Godliness) as a Feature of Character and Religious Education. *The Social Sciences*, 12(6), 1106–1111.

[21] Huda, M., Siregar, M., Ramlan, Rahman, S.K.A., Mat Teh, K.S., Said, H., Jamsari, E.A., Yacub, J., Dacholfany, M.I. and Ninsiana, W. (2017j). From Live Interaction to Virtual Interaction: An Exposure on the Moral Engagement in the Digital Era. *Journal of Theoretical and Applied Information Technology*, 95(19), 4964–4972.

[22] Huda, M., Maseleno, A., Jasmi, K.A., Mustari, I. and Basiron, B. (2017k). Strengthening Interaction from Direct to Virtual Basis: Insights from Ethical and Professional Empowerment. *International Journal of Applied Engineering Research*, 12(17), 6901–6909.

[23] Huda, M., Haron, Z., Ripin, M.N., Hehsan, A. and Yaacob, A.B.C. (2017l). Exploring Innovative Learning Environment (ILE): Big Data Era. *International Journal of Applied Engineering Research*, 12(17), 6678–6685.

[24] Huda, M. and Teh, K.S.M. (2018). Empowering Professional and Ethical Competence on Reflective Teaching Practice in Digital Era. In K. Dikilitas, E. Mede, and D. Atay (Eds.), *Mentorship Strategies in Teacher Education* (pp. 136–152). Hershey, PA: IGI Global. http://doi.org/10.4018/978-1-5225-4050-2.ch007.

[25] Huda, M., Teh, K.S.M., Nor, N.H.M. and Nor, M.B.M. (2018a). Transmitting Leadership Based Civic Responsibility: Insights from Service Learning. *International Journal of Ethics and Systems*, 34(1), 20–31. http://doi.org/10.1108/IJOES-05-2017-0079.

[26] Huda, M., Maseleno, A., Muhamad, N.H.N., Jasmi, K.A., Ahmad, A., Mustari, M.I. and Basiron, B. (2018b). Big Data Emerging Technology: Insights into Innovative Environment for Online Learning Resources. *International Journal of Emerging Technologies in Learning*, 13(1), 23–36. http://doi.org/10.3991/ijet.v13i01.6990.

[27] Huda, M., Maseleno, A., Teh, K.S.M., Don, A.G., Basiron, B., Jasmi, K.A., Mustari, M.I., Nasir, B.M. and Ahmad, R. (2018c). Understanding Modern Learning Environment (MLE) in Big Data Era. *International Journal of Emerging Technologies in Learning*, 13(5), 71–85. http://doi.org/10.3991/ijet.v13i05.8042.

[28] Huda, M. (2018b). Empowering Application Strategy in the Technology Adoption: Insights from Professional and Ethical Engagement. *Journal of Science and Technology Policy Management*, doi.org/10.1108/JSTPM-09-2017-0044.

[29] Huda., M. and Sabani, N. (2018). Empowering Muslim Children's Spirituality in Malay Archipelago: Integration between National Philosophical Foundations and Tawakkul (Trust in God). *International Journal of Children's Spirituality*, 23(1), 81–94.

[30] Huda, M., Almunawar, M.N., Hananto, A.L., Rismayadi, B., Jasmi, K.A., Basiron, B. and Mustari, M.I. (2018). Strengthening Quality Initiative for Organization Stability: Insights from Trust in Cyberspace-Based Information Quality. In *Cases on Quality Initiatives for Organizational Longevity* (pp. 140–169). Hershey, PA: IGI Global. http://doi.org/10.4018/978-1-5225-5288-8.ch006.

[31] Huda, M., Qodriah, S.L., Rismayadi, B., Hananto, A., Kardiyati, E.N., Ruskam, A. and Nasir, B.M. (2019). *Towards Cooperative with Competitive Alliance: Insights into Performance Value in Social Entrepreneurship in Creating Business Value and Competitive Advantage with Social Entrepreneurship* (p. 294). Hershey, PA: IGI Global. http://doi.org/10.4018/978-1-5225-5687-9.ch014.

[32] Huda, M., Hehsan, A., Basuki, S., Rismayadi, B., Jasmi, K.A., Basiron, B. and Mustari, M.I. (2019). Empowering Technology Use to Promote Virtual Violence Prevention in Higher Education Context. In *Intimacy and Developing Personal Relationships in the Virtual World* (pp. 272–291). Hershey, PA: IGI Global. http://doi.org/10.4018/978-1-5225-4047-2.ch015.

[33] Huda, M., Ulfatmi, Luthfi, M.J., Jasmi, K.A., Basiron, B., Mustari, M.I., Safar, A., Embong, H.W.H., Mohamad, A.M. and Mohamed, A.K. (2019). *Adaptive Online Learning Technology: Trends in Big Data Era in Diverse Learning Opportunities Through Technology-Based Curriculum Design* (pp. 163–195), Hershey, PA: IGI Global. http://doi.org/10.4018/978-1-5225-5519-3.ch008.

[34] Huda, M., Muhamad, N.H.N., Teh, K.S.M., Don, A.G., Mulyadi, D. and Hananto, A.H. (2018). Empowering Corporate Social Responsibility (CSR): Insights from Service Learning. *Social Responsibility Journal.* (In Press)

[35] Kartanegara, M. and Huda, M. (2016). Constructing Civil Society: An Islamic Cultural Perspective. *Mediterranean Journal of Social Science*, 7(1), 126–135.

[36] Kurniasih, D., Jasmi, K.A., Basiron, B., Huda, M., Maseleno, A. (2018). The Uses of Fuzzy Logic Method for Finding Agriculture and Livestock Value of Potential Village. *International Journal of Engineering & Technology*, 7(3), 1091–1095.

[37] Maseleno, A., Pardimin, Huda, M., Ramlan, Hehsan, A., Yusof, Y.M., Haron, Z., Ripin, M.N., Nor, N.H.M. and Junaidi, J. (2018a). Mathematical Theory of Evidence to Subject Expertise Diagnostic. *ICIC Express Letters*, 12(4), 369 http://doi.org/10.24507/icicel.12.04.369.

[38] Maseleno, A., Huda, M., Jasmi, K.A., Basiron, B., Mustari, I., Don, A. G. and Ahmad, R. (2018b). Hau-Kashyap Approach for Student's Level of Expertise. *Egyptian Informatics Journal*, DOI.org/10.1016/j.eij.2018.04.001.

[39] Maseleno, A., Sabani, N., Huda, M., Ahmad, R., Jasmi, K.A., Basiron, B. (2018c). Demystifying Learning Analytics in Personalised Learning. *International Journal of Engineering & Technology*, 7(3). 1124–1129.

[40] Maseleno, A., Huda, M., Siregar, M., Ahmad, R., Hehsan, A., Haron, Z., Ripin, M.N., Ihwani, S. S. and Jasmi, K.A. (2017). Combining the Previous Measure of Evidence to Educational Entrance Examination. *Journal of Artificial Intelligence*, 10(3), 85–90.

[41] Moksin, A.I., Shahrill, M., Anshari, M., Huda, M. and Tengah, K.A. (2018b). The Learning of Integration in Calculus Using the Autograph Technology. *Advanced Science Letters*, 24(1), 550–552.

[42] Aminudin, N., Huda, M., Ihwani, S.S., Mohd Noor, S.S., Basiron, B., Jasmi, K.A., Safar, J., Mohamed, A.K., Wan Embong, Wan H., Mohamad, A.M., Maseleno, A., Masrur, M., Trisnawati and Rohmadi, D. (2018). The Family Hope Program using AHP Method. *International Journal of Engineering and Technology*, 7(2.27), 188–193.

[43] Aminudin, N., Huda, M., Kilani, A., Wan Embong, W.H., Mohamed, A.M., Basiron, B., Ihwani, S.S., Sulaiman Shakib Mohd Noor, Kamarul Azmi Jasmi, Jimaain Safar, Natalie L. Ivanova, Andino Maseleno, Agus Triono and Nungsiati. (2018). Higher Education Selection using Simple Additive Weighting. *International Journal of Engineering and Technology*, 7(2.27), 211–217.

[44] Aminudin, N. F., Huda, M., Hehsan, A., Ripin, Mohd. N., Haron, Z., Junaidi, J., Irviani, R., Muslihudin, M., Hidayat, S., Maseleno, A., Gumanti, M. and Fauzi, A. N. (2018). Application Program Learning Based on Android for Students Experiences. *International Journal of Engineering and Technology*, 7(2.27), 194–198.

[45] Othman, R., Shahrill, M., Mundia, L., Tan, A. and Huda, M. (2016). Investigating the Relationship Between the Student's Ability and Learning Preferences: Evidence from Year 7 Mathematics Students. *The New Educational Review*, 44(2), 125–138.

[46] Pardimin, A., Ninsiana, W., Dacholfany, M. I., Kamar, K., Mat Teh, K. S., Huda, M., Hananto, A. L., Muslihudin, M., K. Shankar and Maseleno, A. (2018). Developing Multimedia Application Model for Basic Mathematics Learning. *Journal of Advanced Research in Dynamical and Control Systems*. (in press).

[47] Putra, D.A.D., Jasmi, K.A., Basiron, B., Huda, M., Maseleno, A., Shankar, K. and Aminudin, N. (2018). Tactical Steps for E-Government Development. International *Journal of Pure and Applied Mathematics*, 119(15), 2251–2258.

[48] Ristiani, P., Mat Teh, K. S., Fauzi, A., Hananto, A. L., Huda, M., Muslihudin, M., Shankar, K. and Maseleno, A. (2018). Decision Support System Model for Selection of Best Formula Milk for Toddlers Using Fuzzy Multiple Attribute Decision Making. *Journal of Advanced Research in Dynamical and Control Systems*. (in press).

[49] Rosli, M.R.B., Salamon, H. B. and Huda, M. (2018). Distribution Management of Zakat Fund: Recommended Proposal for Asnaf Riqab in Malaysia. *International Journal of Civil Engineering and Technology*, 9(3), 56–64.

[50] Abadi, S., Mat Teh, K. S., Nasir, B. M., Huda, M., Ivanova, N. L. Sari, T. I., Maseleno, A., Satria, F. and Muslihudin, M. (2018). Application Model of k-means Clustering: Insights into Promotion Strategy of Vocational High School. *International Journal of Engineering and Technology*, 7(2.27), 182–187.

[51] Abadi, S., Huda, M., Jasmi, K.A., Noor, S.S.M., Safar, J., Mohamed, A.K., Wan Embong, W.H., Mohamad, A.M., Hehsan, A., Basiron, B., Ihwani, S. S., Maseleno, A., Muslihudin, M., Satria, F., Irawan, D. and Hartati, Sri. (2018). Determination of the Best Quail Eggs using Simple Additive Weighting. *International Journal of Engineering and Technology*, 7(2.27), 225–230.

[52] Abadi, S., Huda, M., Hehsan, A., Mohamad, A.M., Basiron, B., Ihwani, S. S., Jasmi, K.A., Safar, J., Mohamed, A.K., Embong, H. W., Noor, S.S.M., Brahmono, B., Maseleno, A., Fauzi, A.N., Aminudin, N. and Gumanti, M. (2018). Design of Online Transaction Model on Traditional Industry in Order to Increase Turnover and Benefits. *International Journal of Engineering and Technology*, 7(2.27), 231–237.

[53] Abadi, S., Huda, M., Basiron, B., Ihwani, S. S., Jasmi, K.A., Hehsan, A., Safar, J., Mohamed, Ahmad K., Embong, W.H.W., Mohamad, A.M., Noor, Sulaiman Shakib Mohd., Novita, D., Maseleno, A., Irviani, R., Idris, M. and Muslihudin, M. (2018). Implementation of Fuzzy Analytical Hierarchy Process on Notebook Selection. *International Journal of Engineering and Technology*, 7(2.27), 238–243.

[54] Abadi, S., Mat Teh, K. S., Huda, M., Hehsan, A., Ripin, M. N., Haron, Z., Muhamad, N.H.N., Rianto, R., Maseleno, A., Renaldo, R. and Syarifudin, A. (2018). Design of Student Score Application for Assessing the Most Outstanding Student at Vocational High School. *International Journal of Engineering and Technology*, 7(2.27), 172–177.

[55] Sugiyarti, E., Jasmi, K.A., Basiron, B., Huda, M., Shankar, K. and Maseleno, A. (2018). Decision Support System of Scholarship Grantee Selection using Data Mining. *International Journal of Pure and Applied Mathematics*, 119(15), 2239–2249.

[56] Sundari, E., Jasmi, K.A., Basiron, B., Huda, M. and Maseleno, A. (2018). Web-Based Decision Making System for Assessment of Employee Revenue using Weighted Product. *International Journal of Engineering and Technology*.

[57] Susilowati, T., Jasmi, K.A., Basiron, B., Huda, M., Shankar, K., Maseleno, A., Julia, A. and Sucipto. (2018). Determination of Scholarship Recipients Using Simple Additive Weighting Method. *International Journal of Pure and Applied Mathematics*, 119(15), 2231–2238.

[58] Susilowati, T., Ihsan Dacholfany, M., Aminin, S., Ikhwan, A., Nasir, B.M., Huda, M., Prasetyo, A., Maseleno, A., Satria, F., Hartati, Sri and Wulandari. (2018). Getting Parents Involved in Child's School: Using Attendance Application System Based on SMS Gateway. *International Journal of Engineering and Technology*, 7(2.27), 167–174.

[59] Susilowati, T., Mat Teh, K.S., Nasir, B.M., Don, A.G., Huda, M., Hensafitri, T., Maseleno, A., Oktafianto and Irawan, D. (2018). Learning Application of Lampung Language Based on Multimedia Software. *International Journal of Engineering and Technology*, 7(2.27), 175–181.

[60] Wulandari, S.A., Ihsan Dacholfany, M., Mujib, A., Huda, M., Nasir, B.M., Maseleno, A., Sundari, E., Fauzil and Masrur, M. (2018). Design of Library Application System. *International Journal of Engineering and Technology*, 7(2.27), 199–204.

[61] Zakirillah, Noorminshah, A.I., Huda, M., Fathoni and Heroza, R.I. (2016). Design of a Mobile based Academic Cyber Counselling Application in Higher Education. *Journal of Information Systems Research and Innovation*, 10(3), 1–9.

[62] Septiropa, Z., Osman, Mohd. H., Rahman, A.B.A., Ariffin, M.A.M., Huda, M. and Maseleno, A. (2018). Profile of Cold-Formed Steel for Compression Member Design a Basic Combination Performance. *International Journal of Engineering and Technology*, 7(2.27), 284–290.

[63] T. Avudaiappan, R. Balasubramanian, S. Sundara Pandiyan, M. Saravanan, S.K. Lakshmanaprabu and K. Shankar (2018). Medical Image Security Using Dual Encryption with Oppositional Based Optimization Algorithm. *Journal of Medical Systems-Springer*, September 2018. DOI: https://doi.org/10.1007/s10916-018-1053-z.

[64] Karthikeyan, K., Sunder, R., Shankar, K., Lakshmanaprabu, S.K., Vijayakumar, V. Elhoseny, M. and Manogaran, G. (2018). Energy Consumption Analysis of Virtual Machine Migration in Cloud using Hybrid Swarm Optimization (ABC–BA). *The Journal of Supercomputing*, September 2018. DOI: https://doi.org/10.1007/s11227-018-2583-3.

[65] Shankar, K., Lakshmanaprabu, S.K., Gupta, D., Maseleno, A. and de Albuquerque, V.H.C. (2018). Optimal Feature-based Multi-kernel SVM Approach for Thyroid Disease Classification. *The Journal of Supercomputing*, 1–16.

[66] Shankar, K., Elhoseny, M., Lakshmanaprabu, S.K., Ilayaraja, M., Vidhyavathi, R.M. and Alkhambashi, M. (2018).Optimal Feature Level Fusion Based ANFIS Classifier for Brain MRI Image Classification. *Concurrency and Computation: Practice and Experience*, June 2018. DOI: https://doi.org/10.1002/cpe.4887.

[67] Lakshmanaprabu, S.K., Shankar, K., Khanna, A., Gupta, D., Rodrigues, J.J., Pinheiro, P.R. and De Albuquerque, V.H.C. (2018). Effective Features to Classify Big Data Using Social Internet of Things. *IEEE Access*, 6, 24196–24204.

[68] Shankar, K. and Eswaran, P. (2016). RGB-Based Secure Share Creation in Visual Cryptography Using Optimal Elliptic Curve Cryptography Technique. *Journal of Circuits, Systems and Computers*, 25(11), 1650138.

[69] Shankar, K. and Eswaran, P. (2015). A Secure Visual Secret Share (VSS) Creation Scheme in Visual Cryptography using Elliptic Curve Cryptography with Optimization Technique. *Australian Journal of Basic & Applied Science*, 9(36), 150–163.

[70] Shankar, K. and Eswaran, P. (2015). ECC Based Image Encryption Scheme with Aid of Optimization Technique using Differential Evolution Algorithm. *International Journal of Applied Engineering Research,* 10(55), 1841–1845.

[71] Rizk-Allah, R. M., Hassanien, A. E., Elhoseny, M. and Gunasekaran, M. A New Binary Salp Swarm Algorithm: Development and Application for Optimization Tasks. *Neural Computing and Applications*, 1–23.

[72] Gharbia, R., Hassanien, A. E., El-Baz, A. H., Elhoseny, M. and Gunasekaran, M. (2018). Multi-Spectral and Panchromatic Image Fusion Approach using Stationary Wavelet Transform and Swarm Flower Pollination Optimization for Remote Sensing Applications. *Future Generation Computer Systems*.

[73] Hosseinabadi, A.A.R., Vahidi, J., Saemi, B., Sangaiah, A.K. and Elhoseny, M. (2018). Extended Genetic Algorithm for Solving Open-shop Scheduling Problem. *Soft Computing*, Springer, April 2018 (https://doi.org/10.1007/s00500-018-3177-y)

[74] Hassanien, A. E., Rizk-Allah, R. M. and Elhoseny, M. (2018). A Hybrid Crow Search Algorithm Based on Rough Searching Scheme for Solving Engineering Optimization Problems. *Journal of Ambient Intelligence and Humanized Computing*, 1–25.

[75] Elhoseny, M., Yuan, X., Yu, Z., Mao, C., El-Minir, H. and Riad, A. (2015). Balancing Energy Consumption in Heterogeneous Wireless Sensor Networks using Genetic Algorithm. *IEEE Communications Letters, IEEE*, 19(12): 2194–2197, http://doi.org/10.1109/LCOMM.2014.2381226.

[76] Gafar, M. G., Elhoseny, M. and Gunasekaran, M. (2018). Modeling Neutrosophic Variables Based on Particle Swarm Optimization and Information Theory Measures for Forest Fires. *The Journal of Supercomputing*, 1–18.

[77] Metawa, N., Hassan, M. K. and Elhoseny, M. (2017). Genetic Algorithm Based Model for Optimizing Bank Lending Decisions. *Expert Systems with Applications*, 80, 75–82.

8 The investment opportunities for building smartphone applications for tourist cities in Saudi Arabia

The case of Abha City

Saeed Q. Al-Khalidi Al-Maliki and Mohammed A. Al-Ghobiri

Introduction

Since its advent, the internet has provided many services to humanity and continues to provide an opportunity for the exchange of information between people throughout the world. The internet has led to more possibilities, facilities, and opportunities for research, education, health, commerce, and other sectors. Therefore information and communication technology (ICT) has become a way of life, not simply a luxury tool for a particular field or social elites. In light of the global trend towards knowledge economies, which rely mainly on modern technologies to gain knowledge in raising social welfare and exploiting various resources, ICT has become a means of survival and an indispensable tool in an open and competitive world as well as a criterion for progress and prosperity.

Accelerated technological development in the mid-1990s led to a boom at all educational levels and the launch of a real revolution in the world of communication. The internet has spread around the world, making the planet into a small village, opening societies to each other, and making it easier to exchange views, ideas, and experiences. The internet is now the best means of communication between individuals and groups.

By using ICT tools and applications, the internet, and mobile devices to support good governance, government agencies and department initiatives can strengthen existing relationships and build new partnerships within civil society. These often take the form of smartphone applications used by smart mobile initiatives. The main purpose of smartphone applications is to build communication with others. In addition, Vijoli and Marinescu [1] mention that most mobile applications resemble travel guides, digital maps, smart cards, and localisation maps. Mobile applications offer the possibility of booking the accommodation and the flight and finding out the most recommended places to eat out or the activities that can be performed in the respective tourist destination.

Ibrahim and Fawzi [2] state that the Arab countries are weak in their investment in ICT infrastructure. As a result, generally, their tourism offers lack credibility.

In Saudi Arabia, many Saudi youths have taken steps towards building their own smartphone applications to deal with local activities through the internet. It is known that the differing operating systems on mobiles vary between Android, iOS, and Windows. Each has its own independent applications and e-shop, which is the largest platform for launching applications of this type for such operating systems. For example, iOS has the App Store and Android has Google Play. One should, before entering into the world of smartphone applications, determine the most important platforms on which a company would launch its application, as the programming method for the application must comply with the various operating systems. The services offered by these smartphone applications contribute to the competition of many companies, and experts in the production of modern applications with different designs and forms can attract users to download and then buy and sell through them. This means financial profit for them and those in charge of the industry.

The main objective of this paper is to identify the investment opportunities of smartphone applications for Saudi tourist cities, in particular, Abha City. To this end, the results of this study were analysed to determine the point of view of the sample mentioned in the study methodology.

This paper sought to gather the opinions of the participants through interviews by asking them questions related to investment opportunities in smartphone applications for Abha City as the study area. Abha City is a tourist city with multiple tourist elements and was the capital of Arab tourism for the year 2017. The researchers suggest that there are important opportunities that may be a source of livelihood. This is related to the field of investment and the increase in Saudi youth interested in the creation and design of smart applications, who are adapting the design and programming of smart applications on smart devices to the field of tourism. The researcher hopes that this study will increase awareness of the importance of these opportunities in investing in smartphone applications in Saudi tourist cities.

In this research paper, the background study will be presented following the introduction. Then the research problem statement will be discussed, and a research objective and methodology will be presented, followed by the research case study, 'Abha City, Saudi Arabia'. The study questions are then presented, followed by the findings and discussion. Implications and recommendations are offered next, with conclusions presented in the last section.

Background study

Al-Maliki [3] discussed the adoption of ICT as a response to growth discrepancies in Saudi Arabia in particular. The objective of this research study is to identify the impact of ICT investment in Saudi Arabia and the role of the government. Another study by Al-Maliki [4] determined the factors affecting e-governance adoption in Saudi Arabia. It was found that the main barriers were a lack of skills and human resources, low computer literacy and training capacity, and English language difficulties. The researcher proposed a conceptual architecture for e-portal services.

The proposed recommendations addressed the need to understand the adoption of e-government and to help citizens use the available services. The proposed model reiterated that citizens should be helped until they fully understand the functioning of e-government applications.

Ibrahim and Fawzi [2] mentioned in their study that the increase in global tourism demand is due to several factors, including the increase in income in many tourist-exporting countries. This demand has become more important in the field of ICTs, which lead to good communication between the countries exporting tourism and the receiving countries and to the provision of tourist services and hotels that are distinctive for tourists. This technology has become an important factor in pushing the tourism sector and development.

A study by Krishan et al. [5] titled 'Digital Tourism Applications and Their Role in Promoting Digitalisation of Societies and Transformation towards Smart Tourism' aims to highlight the importance of digitising transport and communication services to support tourist cities by providing applications that allow the tourist to request a specific means of transportation within the site. This allows a focus on enabling the system to position services so that the tourists can locate or obtain the nearest means of transport without delay. It therefore speaks of the importance of planning transport and communication systems with consideration of sustainable development and their impact on communities from other sources.

Alshattnawi [6] also notes that the rapid proliferation of mobile computing technology has considerable potential for providing access to different services at any time and from anywhere. It allows users to access several applications and services via an internet connection or by building stand-alone applications. Alshattnawi [6] highlights that the existing tourist guide applications can use the latest technologies to enhance the application quality by satisfying the user's requirements. In her study, Alshattnawi states that several development platforms for mobile applications are used to design tourist guide applications. The main aim of her study was to build electronic tourist guide systems using the two technologies to confirm the power of Android. The applications are tailored to user preferences so that the user can access the application from a simple interface or automatically; the information may be displayed according to a global positioning system (GPS). The system is built to be a tourist guide for Jordan, but it is flexible and easy to use for any country.

In addition, Wang et al. [7] studied smartphone applications and how they support a wide range of information services at any time and from anywhere. Smartphone applications have the potential to significantly influence the touristic experience. This study explores the mediation mechanisms of smartphones by examining stories provided by travellers related to their use of smartphones (and associated applications) for travelling purposes. The results reveal that smartphones can change tourists' behaviour and emotional states by addressing a wide variety of information needs. In particular, the instant information support of smartphones enables tourists to more effectively solve problems, share experiences, and 'store' memories.

Dickinson and colleagues [8] studied the relationship between tourism and smartphone applications. They note that the smartphone has rapidly been adopted as a tourism travel tool. With a growing number of users and a wide variety of applications emerging, the smartphone is fundamentally altering our current use and understanding of the transport network and tourism travel. Based on a review of smartphone apps, their article evaluates the current functionalities used in the domestic tourism travel domain and highlights where the next major developments lie. Their article also analyses how the smartphone mediates tourism travel and the role that it may play in more collaborative and dynamic travel decisions to facilitate sustainable travel.

Mang and colleagues [9] explored smartphone utilisation by tourists from 24 countries visiting Rome, Italy, or Athens, Greece. By extending a standard technology acceptance model, we can identify common travel uses for smartphones, which include taking photos, social networking, viewing maps, finding transportation, and searching for shops and restaurants. Younger cohorts utilise their phones more than older cohorts, but there is no difference in utilisation between females and males. The most important factors affecting behaviour are how often the tourist normally utilises their smartphone when at home and whether or not the tourist has non-WiFi data access.

In a study by Lin and colleagues [10], the main purpose was to develop an application (or 'app') by integrating GPS to provide digitised information of local cultural spots to guide tourists for tourism promotion and the digitised information of mountaineering trails to monitor energy expenditure (EE) for health promotion. The provided cultural information was also adopted for educational purposes. The Extended Technology Acceptance Model (TAM) was used to evaluate the usefulness and behaviour intention of the provided information and functions in the developed system. Most users agreed that the system is useful for health promotion, tourism promotion, and folk-culture education. They also showed strong intention and a positive attitude towards continuous use of the app.

Gretze and colleagues [11] state that 'smart tourism' is a new buzzword applied to describe the increasing reliance of tourism destinations, their industries, and their tourists on emerging forms of ICT that allow for large amounts of data to be transformed into value propositions. However, it remains ill-defined as a concept, which hinders its theoretical development. The present paper defines smart tourism, sheds light on current smart tourism trends, and then lays out its technological and business foundations. This is followed by a brief discussion on the prospects and drawbacks of smart tourism. The paper further draws attention to the great need for research to inform smart tourism and its development and management.

Mihajlović [12] seeks to understand the mission and the usage of ICT through its influence on the development of the supply of tourism subjects, especially intermediaries in the tourism market. The occurrence of the phenomenon of tourism is the result of socio-economic conditions related to the development of technology and innovations. The first part of the paper gives an overview of the theoretical insights that complement the argument regarding the role of technology in tourism. The second part of this paper, based on the results of a survey that was conducted in Croatian travel agencies, explores the role of technology in the business of travel agencies through the reviews of managers. Results of the research indicate the new trends in activities of travel agencies that influence the development tendencies of intermediaries.

Mateia [13] studied the importance of the internet for the travel and tourism industry, which has increased rapidly over the last few years, because the new technology allows both active access to information and increasing coverage. For this reason, tour operators have recently developed applications that can be found in mobile web applications.

Sava and Mateia [14] state that development of tourism is correlated with new means of communication and information that reaches any destination around the world instantaneously. Travel agencies that represent an intermediate agent between tourist services and potential tourists are threatened by the continuous development of communication technologies, as it is difficult to keep up with the new trends. Thus applications must be developed for smartphones to become more attractive and useful for potential tourists who must be informed in real time about destinations.

Research problem statement

Studies on smart mobile phone applications have been very limited, with the exception of a small number of daily newspaper reports and investigations. This research paper attempts to study the investment opportunities for smart applications and to identify the main challenges that may hinder successful construction of smart applications. Al-Maliki [3] indicates that literacy barriers, technical factors, and lack of awareness and trust could be the main challenges of developing this type of business.

A review of the literature has revealed that there is limited empirical research on the awareness of building smart applications used by internet phone devices from citizens' perspectives. According to Al-Maliki [3], cultural differences can result in limitations in ICT implementation. For example, different views on logic and reasoning and deficiencies in language use can create barriers to effective communication and understanding.

Regarding the ICT investment in Saudi Arabia, a significant market for networking and related software, as well as for computer hardware and software products, has developed [15]. Saudi Arabia is currently the largest computer market in the Gulf region.

Al-Maliki [16] found that Saudi Arabia's overall ICT spending hit $5.7 billion by the end of 2014, up from around $3.5 billion in 2010, with a compound annual growth rate of around 13%. This huge expenditure on its ICT sector includes projects such as an e-government portal, e-government network, e-government interoperability framework, e-payment gateway (Sadad), e-tax system, and electronic information exchange. According to Aleqtisadiah [17], the total spending on communications and information technology (IT) services in Saudi Arabia, including services, hardware, and software, reached $223 billion in 2015, which represented growth of 150% since 2006.

As a result of ICT investment in Saudi Arabia, many government services, such as those for investment licenses, visa applications, traffic ticket enquiries and payment, paying passport fees, and paying utility bills, are currently available online.

This research is assuming that ICT applications have an important role in driving the growth of tourism within Saudi Arabia, particularly in the city of Abha. Growth in ICT would highly improve communications with tourists and provide them with essential tourist information and convenient tools for managing their tourist journey.

Research objectives and methodology

The main objectives of this paper are to evaluate the extent of investment opportunities among Saudi youths in building smart applications. It also aims to identify the main factors that may inhibit attitudes towards building smartphone applications.

The research adopted a descriptive and analytical approach for the study of building smartphone applications. This research approach was considered appropriate for analysing Saudi youths' awareness of the investment opportunities for building smartphone applications in Saudi Arabia, and Abha City in particular as a tourist city.

A literature review was also conducted on this topic. Information was collected from articles published by other researchers and from current trends in this area.

The research paper considers some aspects of the investment opportunities for building smartphone applications. Many issues are covered, including the extent of awareness about the use of smartphone applications in the Saudi context and the main factors hindering building smartphone applications over the internet. The study also covers the main reasons why Saudis have not been enthusiastic towards this type of business, particularly in Abha City. The main factors that could lead to resistance or failure to build smartphone applications are covered. In addition, the study will suggest solutions to the lack of awareness among Saudi youths in building smartphone applications.

The study sample consisted of the Faculty of Administrative and Financial Sciences students at King Khalid University in Abha, particularly students undertaking the course of Management Information Systems (MIS), which consists of four divisions with 282 students. The students were part of four faculty departments: Business Administration, Accounting, Electronic Commerce, and Law. The eight survey questions were asked through the Blackboard e-learning program. A total of 240 students responded to the study questionnaire, representing 85% of the sample, which was considered sufficient for this research. The researchers requested that students express their opinions about the study issue by answering the eight questions related to the investment opportunities for building smartphone applications for tourist cities in Saudi Arabia, specifically in the case of Abha City.

Abha City, Saudi Arabia

Background

Abha, the home of the headquarters of the regional governorate and capital of the Asir region, is located in the south-west of the Kingdom. Abha is a well-known hill station in the Kingdom of Saudi Arabia. It is mainly known for its cold climate during hot summers, ample rainfall, and green landscape compared to the hot, dry, and barren desert in the majority of the Kingdom area. The city is situated at an elevation of 2,270 m (7,500 feet) above sea level and is a mountain retreat and vacation spot for people from across Saudi Arabia and the other Arabian Gulf countries due to its relatively moderate climate. It has a diverse environment with forests in Soudah, Dulghan, Al-Faraa, and so forth. There are also towering highlands in addition to the eastern plains and western coasts.

The town has a population of about 450,000 [18]. Abha is one of the top destinations for tourists in the Kingdom of Saudi Arabia, which has resulted in exponential growth in development in terms of buildings, roads, and so forth.

Business opportunities in Abha City

Abha has been named the 'Arab Capital of Tourism' for 2017, which opens the door for various business opportunities. Kalaiya and Kumar [19] state:

> Tourism is a collection of activities, service and industries which deliver a travel experience comprising transportation, accommodation, eating and drinking establishments, retail shops, entertainment, business and other hospitality services provided for individual or group traveling away from home.

The study questions

To achieve the study objectives, literature on the awareness of using e-services and building smart applications among Saudi youths has been reviewed. Further, the following questions on this topic will be addressed:

1 Why has the trend moved towards the use of mobile applications as a means of effective marketing in Saudi Arabia?
2 What are the principal reasons behind Saudi citizens investing in building smartphone applications?
3 How is investment done in mobile applications?
4 How can you make the most of mobile applications?
5 Is it financially an encouraging business?
6 How can we advertise in smartphone applications?
7 What type of investment opportunities for building smartphone applications are available to serve Abha as a tourist city?
8 What are the obstacles to investment in smartphone applications?

Findings and discussion

The study indicates that there are investment opportunities for building smartphone applications. In some cases, Saudi youths in Abha City do not understand how to take advantage of investment opportunities for building smartphone applications due to a lack of awareness. In Abha City, relevant public and private tourist agencies should provide support and help for anyone with suggestions for investment opportunities for smartphone applications.

The researchers hope that Saudi youths will become interested in this field, thus providing the appropriate environment and necessary support for the production of such applications, as they have a promising economically productive future. The development of smartphone applications has transformed from a hobby into a source of income, generating thousands of riyals. The market has experienced remarkable demand, as millions of users seek to purchase and download mobile applications of various kinds. In the following paragraphs, the researchers have analysed the answers to the survey questions that were addressed above.

Why has the trend moved towards the use of mobile applications as a means of effective marketing in Saudi Arabia?

In relation to this question, mobile applications are considered among the most powerful marketing tactics of companies today. Phone applications began to emerge for the first time at the end of the 1990s, though they were limited to a number of primitive programs, such as games, a calculator, and a calendar. With the smartphone revolution, a similar revolution has occurred in the development and manufacturing of applications. This has made the experience of using the phone more enjoyable and useful, and, in turn, the phone has become an essential part of our daily lives. The increasing reliance on smartphones has pushed many young Saudis to compete in developing applications that meet basic needs and enable users to get different services in all areas of life. Moreover, it has met the demands of youth in securing jobs and increasing the economic opportunities available to them. The application development is not very demanding, as the platforms are ready and services exist. A developer only needs to use the codes and software that perform a particular function.

Mobile applications help and support businesses in regard to evolution, diversity, and other benefits that can be obtained from such applications. The existence of a special application for the products and services provides a golden opportunity for the user to see it every time the smartphone is used.

By 2016, the total expenditure on marketing operations through smartphones is expected to reach $2.8 billion in the Middle East alone [20]. The smartphone applications market is thus an effective and irreplaceable electronic marketing tool. Therefore companies seek to keep up with the development of applications for mobile users and their products and services. This could make a substantial difference in the sales of these companies and increase returns significantly.

What are the principal reasons behind Saudi citizens investing in building smartphone applications?

The main reason for this shift is that investment in this market does not require huge financial resources, prompting many young Saudis to interact with the world of applications. This gives rise to a unique youth environment of amateur and professional application developers whose productions have emerged in several Arab and local programs.

How is investment done in mobile applications?

An idea alone is not enough, as you must look at the big picture for these projects even if you do not have the required programming skills to develop that program, application, or service. In order for your digital dreams not to turn into a mirage, you should study the many aspects involved. One of these aspects is the company developing your idea, so you should keep in mind the process of designing iPhone or Android applications, programming iPhone or Android applications, or programming for both iPhone and Android. Your choice of a good company that has a cumulative knowledge repertoire should consider the latest methods and programming languages with a group of professional programmers. This investment comes in light of the evolution of the smartphone market, one of the fastest growing markets in the world. This large growth increases the importance and proportion of the market of these applications in Saudi Arabia. Smartphone applications are widely used in Saudi Arabia, perhaps more than in even some very advanced countries. This contributes to and enhances the investment in this fertile field. However, few Saudi companies are specialising in the field of smartphone applications.

How can you make the most of mobile applications?

Application development starts with the innovation of the idea, developing the idea, and identifying the target group; thereafter, an intensive advertising campaign via the internet and social media can be carried out. In addition, making mobile applications is achieved through the use of professional electronic applications for mobiles, so as to provide the user ease of use and comfort in operation and at the same time maintain the required benefit effectively and efficiently. Mobile applications can be created either by hiring a mobile application developer or entrusting this task to a specialised company that provides you with all the facilities that you require in your application so that it will appear professional and interesting. Therefore, if any Saudi youths wish to invest in the field of phone applications, they must generate a creative and innovative idea and then head to one of the sites that help in the building of the application in order to upload it later to known download platforms.

Is it financially an encouraging business?

It is true that such projects do not provide opportunities for employment and competition is high. Every year, a new application gains the trust of users within a short period. However, it is a promising field and an indication of the great success that applications could achieve. As for the financial returns, the application may be free, but the advertising spaces within it or the special features within the application can be sold. However, the application can remain free to download.

How can we advertise in smartphone applications?

Start marketing your application on various social networking channels such as Facebook, Twitter, and Instagram, and spare no effort in doing so. Simply, as we mentioned earlier, as soon as your potential customer downloads your company application, it is highly possible for him or her to keep up with the application until the customer becomes familiar with it and it becomes a part of his or her daily practices. Media campaigns are divided into two main parts: preparation and execution. The most difficult is the preparation stage, because it establishes executive sub-plans for a campaign's plan that may last for months, weeks, or days.

The implementation phase is also important in terms of the accuracy and the professionalism of the team performing the steps, as well as the development of new trends in line with the movements of the market and the competition. It is important to examine the means of advertising and knowledge of advertising tools by estimating the social class and age that you will be dealing with primarily. Marketing through the internet has proven its importance in accessing and influencing the largest class of young people. Therefore, many institutions and media bodies use it for promotions and publicity. Many of the ideas and projects that have turned into global sites and large institutions stand as the strongest evidence of the interest in online advertising campaigns and promotions.

Smartphone applications can be downloaded from sites such as Google Play and the App Store. These applications are some of the most important services that phones provide to users.

The benefit of these applications lies in the nature and quality of services offered. The tendency observed in the smartphone applications industry market has led to a large investment revolution. At the same time, the growth of the technical volume of investment in this area has been noted by companies, capitals, and highly technical experts, while others consider it as an additional source of income. It is worth mentioning that the process of investing in such applications comes through their production, reselling, or investment via advertisements.

What type of investment opportunities for building smartphone applications are available to serve Abha as a tourist city?

The types of applications that can be designed exceed the categories of the app stores, which currently include such categories as food, sports, and lifestyle. This type of application is ideal for areas of business that need a specialised application to view information, and benefits include the ease and simplicity of implementation.

Getting a mobile application for your business can serve as a new engine for the success of your marketing task.

Some applications that can be designed for Abha City as a tourist city are as follows:

- Tourist routes
- Popular parks and resorts
- Popular tourist destinations
- Popular cuisine
- Popular festivals
- Popular religious or traditional events
- Reservations of furnished apartments
- Tourist agency (coordinates the shelters and tourists)
- Delivery of food orders to homes
- Natural products such as honey, fruits, and vegetables
- Provision of travel supplies
- Households
- Private lessons
- School and private transport
- Domestic workers
- Textbooks and study abstracts (specialty shops)
- Property
- Publishing the most important tourist activities in the tourist seasons
- Preparation of tourism programs for tracks
- Productive families
- Museums and libraries.

What are the obstacles to investment in smartphone applications?

The obstacles that owners may face in developing and implementing smartphone applications include:

- The owner may not be familiar with these applications' approaches
- Methods of calling
- Lack of the smartphone application's feasibility studies
- Lack of sponsorships for these applications
- Fear of failure in developing and implementing a smartphone application.

Implications

The main findings of this study indicate that smartphone applications contribute positively to raising the quality of e-service awareness among users. They also encourage youths to use these e-services and participate in building new smartphone applications. However, relevant government tourist agencies should provide enough help and support for young Saudi citizens in this regard.

The findings provide indicators to help public and private tourist agencies in Abha to improve their understanding of how to encourage Saudi youths in Abha to build smartphone applications, particularly tourist services applications. This will encourage the public to invest in this modern technology. In addition, increased awareness of building smartphone applications will raise the income of companies and Saudi youths.

Furthermore, this awareness will play a major role in overcoming some constraints of investment in smartphone applications through the help of local private sponsors and governmental agencies. This could include King Khalid University, the Tourism Authority represented by the Tourism Committee and the Tourism Authority in Abha, the Chamber of Commerce in Abha, and the Abha Award for Excellence. These tourist agencies in Abha City should prioritise the assistance and support of Saudi youths, in order to build reliable smartphone applications. This requires special courses to educate developers on how to build their applications in a way to provide reliable services and achieve a good income. In addition, various competitions should be held to motivate young people to innovate and engage with this business. The introduction of relevant subjects in a school's curriculum is also highly recommended. Consequently, this would contribute positively to the creation of appropriate smartphone applications that could enhance various activities locally and globally.

Recommendations

The main objective of this paper research was to explore investment opportunities with regards to the smartphone applications.

After reviewing the main findings of the study, it is now appropriate to consider some recommendations. Having conducted this study in the field of investment opportunities in smartphone applications, it is necessary to identify the most important opportunities that could be used to design smartphone applications that help to connect tourism and tourist services in Abha City.

Investment in smartphone applications will benefit tourism activities in tourist cities, especially in Abha City. This study indicates that there are several important areas of tourism related to investment opportunities in smartphone applications.

In order to gain the benefits of creating smartphone applications, there must be awareness among Saudi youth about how to identify ideas about designing smartphone applications. There is a clear need to raise awareness about the importance of smart applications and their design and programming methods.

The official tourist agency in Abha must encourage young people to contribute to tourism activities by creating ideas about smartphone applications and how to utilise them for tourist activities in Abha City. The researcher recommends developing an experience office to help young people to create smartphone applications. This will encourage the creation and use of smart applications in tourism. This research presents the findings and recommendations that can help Saudi youth to learn about investment opportunities in smartphone applications. The findings will further assist Saudi youth to understand their current situation and identify the benefits of these applications, as this can then lead to understanding the benefits of investing in smartphone applications. The present study indicates that tourism in Saudi Arabia is suitable for investment through the creation of smart applications. The state of current awareness may not contribute to the use of technology, especially in the area of smartphone applications. It is clear that many of Abha City's youth are not fully aware of the potential investment opportunities in smart mobile applications. In addition, the presence of tourism is crucial and provides an opportunity for those with experience in this field of smartphone application development. Support and guidance are some of the obstacles to gaining support for Saudi youth from official bodies, and many are unaware of the field of modern technology. In order to obtain the benefits of choosing an idea for smartphone applications, young people must have the appropriate IT background and experience in how to invest in this field, which can be through appropriate IT training programs. Saudi young people must understand modern technology to improve their knowledge of these investment opportunities so that they not only understand the importance of smart applications but also are able to invent the idea, implement it, commercialise it, and then build it in the future. Therefore, Saudi youths must understand the importance of investment in technology, which the current study indicates affects the field of tourism. In addition, the study has identified some factors that are important in helping young people to consider solutions in order to obtain benefits from IT. The idea of smartphone applications requires more attention from those interested in tourism in Abha City. This can be done through programs for Saudi youth who wish to develop their skills and invest in this area to be able to provide better tourist services to the region's tourists. Furthermore, the Tourism Authority in Abha City should encourage the involvement of young people in providing better services for tourism, provide advice, and answer questions on the subject. The ambitious Saudi youth should develop their understanding of the use of modern technology. They can do this by joining programs that have training in investment opportunities in smart mobile applications and IT fields, especially in the field of smartphone applications, in order to increase awareness and learn how to benefit from them, including how they can be applied to tourism services in Saudi tourist cities. In addition, Saudi experts in this regard should consider adopting and supporting young people's ideas through technical bodies such as Makkah Valley Technology and consider the establishment of the Abha Valley Technology in the near future. This will enable the experts to encourage such initiatives and sponsor the expertise offices at King Khalid University in Abha City. Saudi academics and IT professionals should adopt ideas that help young people to develop their own ideas in these areas. Academics and professionals should further provide advice, study youth initiatives, and supervise such initiatives until they are implemented and are serving the relevant tourism fields. This will greatly help in the development and use of information technology in the Kingdom of Saudi Arabia.

It is hoped that the implications of the results of this study and the implementation of the recommendations will encourage relevant initiatives, especially in Abha City. This is because there is a high committee for the development of the Aseer region and its capital Abha City in several areas, including tourism aspects.

Conclusion

Over the last few years, there has been considerable infrastructure development in ICTs in Saudi Arabia. Therefore, efforts must be directed towards the development and implementation of notable ICT services. One important channel in this regard is to encourage investment in smartphone applications for tourist cities, such as the case of Abha City.

The researchers adopted a descriptive and analytical approach, considered appropriate for collecting sufficient qualitative data regarding the investment opportunities for building smartphone applications. The analysis provided a general overview of the current investment opportunities for building smartphone applications. The findings indicate that there is a poor understanding of the investment opportunities for smartphone applications.

This study also showed a lack of smartphone application usage in Abha for tourism, and the lack of support from public and private tourist agencies was evident. This is considered an obstacle for development and usage. There was a definite lack of understanding of how to invest in building smartphone applications. Finally, this study may lead to further in-depth discussion as to the best way to increase awareness of investment opportunities for building smartphone applications among Saudi youths.

References

[1] Vijoli, C. and Marinescu, N. (2016), Analyzing the online promotion of a tourist destination: the case of Saariselkä, *Bulletin of the Transilvania University of Braşov, Series V: Economic Sciences, Finland*, Volume 9 (58), Issue 2, pp. 153–162.

[2] Ibrahim, B. and Fawzi, S. M. (2010), The role of information and communication technology in the tourism and hospitality sector development, *Journal of the Researcher*, Issue 7, pp. 275–285, Retrieved from: http://rcweb.luedld.net/rc7/21-30A2404910.pdf.

[3] Al-Maliki, S. Q. Al-Khalidi, (2013), Information and Communication Technology (ICT) investment in the Kingdom of Saudi Arabia: assessing strengths and weaknesses, *Journal of Organizational Knowledge*, pp. 1–15.

[4] Al-Maliki, S. Q. Al-Khalidi, (2014), Analysis and implementation of factors affecting e-governance adoption in the Kingdom of Saudi Arabia, *International Journal of Strategic Information Technology and Applications*, Volume 5, Issue 1, pp. 20–29.

[5] Krishan, I. Masadeh, R. and Bazazo, T. (2016), Digital tourism applications and their role in promoting digitalization of societies and transformation towards smart tourism, *International Journal of Planning, Urban and Sustainable Development*, Volume 3, Issue 1, Retrieved from: www.researchgate.net/publication/303080460.

[6] Alshattnawi, S. (2013), Building Mobile Tourist Guide Applications using Different Development Mobile Platforms, *International Journal of Advanced Science and Technology*, Volume 1, Issue 54, Retrieved from: www.hindawi.com/journals/mpe/2014/583179/.

[7] Wang, D., Park, S. and Fesenmaier, D. R. (2012), The role of smartphones in mediating the touristic experience, Journal of Travel Research, Volume 51, Issue 4, pp. 371–387.

[8] Dickinson, J, Ghali, K, Cherrett, T, Speed, C, Davies, N and Norgate, S. H. (2012), Tourism and the smartphone app: Capabilities, emerging practice, and scope in the travel domain, *Current Issues in Tourism*, pp. 1–18.

[9]   Mang, C.F., Piper, L.A. and Brown, N.R. (2016), The incidence of smartphone usage among tourists, *International Journal of Tourism Research*, Volume 18, Issue 6, pp. 591–601.

[10] Lin, K.C., Chang, L.S., Tseng, C.M., Lin, H.H., Chen, Y.F. and Chao, C.L. (2014), A smartphone APP for health and tourism promotion, Hindawi Publishing Corporation, *Mathematical Problems in Engineering*, Volume 2014, pp. 1–10.

[11] Gretzel1, U., Sigala, M., Xiang, Z. and Koo, C. (2015), Smart tourism: foundations and developments, *Electron Markets: The International Journal on Networked Business*, Volume 25, pp. 179–188, Retrieved from: https://link.springer.com/article/10.1007/s12525-015-0196-8.

[12] Mihajlović, I. (2012), The impact of Information and Communication Technology (ICT) as a key factor of tourism development on the role of Croatian travel agencies, *International Journal of Business and Social Science*, Volume 3, Issue 24, pp. 151–159, Retrieved from: www.ijbssnet.com/journals/Vol_3_No_24_Special_Issue_December_2012/16.pdf.

[13] Mateia, A.N. (2012), Evolution of tourism applications in virtual media, *Quaestus Multidisciplinary Research Journal*, Volume 03, pp. 158–163, Retrieved from: www.quaestus.ro/wp-content/uploads/2012/03/mateia51.pdf.

[14] Sava, C. and Mateia, A.N. (2012), An analysis of the impact of new communication technologies on tourism, *Quaestus Multidisciplinary Research Journal*, Volume 3, pp. 345–352, Retrieved from: www.quaestus.ro/wp-content/uploads/2012/03/sava.mateia.pdf.

[15] Communications and Information Technology Commission, KSA, (March 2012), E-report No. 10, Retrieved from: www.citc.gov.sa.

[16] Al-Maliki, S.Q. Al-Khalidi, (2016), An exploratory evaluation of the awareness of e-government services among citizens in Saudi Arabia, *Communication, Management and Information Technology, Proceedings of the International Conference on Communication, Management and Information Technology* (ICCMIT 2016, Cosenza, Italy, 26–29 April 2016), CRC Press, pp. 425–432.

[17] Aleqtisadiah (June 2016) Aleqtisadiah. Retrieved from: www.aleqt.com/2016/06/10/article_1061485.html [Accessed 25 February 2018].

[18] Khan, R.A., AlDeen, S., Kafeel, M. and Khan, A.R. (2016), Preliminary investigation of groundwater quality of Abha, Kingdom of Saudi Arabia, *International Journal of Engineering Associates*, Volume 4, Issue 8. Retrieved from: www.advanceresearchlibrary.com/temp/downloads/ijea/August2015/rk7.pdf.

[19] Kalaiya, A.B. and Kumar, A. (2015), Tourism as a development tool: a study on role of tourism in economic development, employment generation and poverty reduction: special focus on Kachchh, *International Journal of Advanced Research in Computer Science and Management Studies*, Volume 3, Issue 7, pp. 189–197. Retrieved from: www.ijarcsms.com/docs/paper/volume3/issue7/V3I7-0073.pdf.

[20] Okaz newspaper official portal, OKAZ. [online] Retrieved from: www.okaz.com.sa/ [Accessed 25 March 2017].

9 An applied credit scoring model

Esther Castro, M. Kabir Hassan and Mark Rosa

Introduction

The banking industry has gone through several changes in the last 60 years. These changes have in part to do with regulatory changes and financial product innovation. Yet one thing has remained: the demand and dominance of consumer lending. Consumer credit loans have increased in the banking industry, in general, as well as in Credit Unions in the last 60 years. Consumer loans have contributed to the way of life for many Americans. For many Americans who wants to increase their standard of living, consumer loans are the answer. Research has shown that consumer loan is among the most profitable loan a bank can make. However, a Functional Cost Analysis (FCA) program conducted by the Federal Reserve found that consumer loans are among the riskiest and costly loanable funds that a bank grants to their customer. Recovering a loan is dependent upon the consumer's economic state, heath state, and many times moral character. Consumer loans, furthermore, are ascertained to be cyclical with the overall state of the economy. With this uncertainty surrounding consumer lending, it poses a challenge for banks to predict loan portfolio risk and loan charge-offs. Charge-off can be defined as an amount of debt that is unlikely to be recovered, thus must be written-off. The recent subprime crises accentuate the need for measuring the portfolio risk of banks. After 2010, bank charge-off in all commercial banks reached $18 billion. Thankfully, this charge-off amount has decreased to about $5 billion. This, however, is still a considerable sum of money capturing the risk for their mortgages, small business loans, or individual borrowers influences the financial institution in making appropriate interest rate, lending policy, and reserve requirement changes.

One major advantage that smaller community banks and credit unions have over large banks is their relationship with customers. A key factor when analyzing a consumer's loan application is to have knowledge of the borrower's character and ability to pay. Knowing the person's character and thus their sense of moral responsibility is a good indication of their intentions to pay back. A consumer loan officer also should seek insight of customer's credit history. There are over 2,000 credit bureaus in the United States that provide credit rating for most individuals who has at one time or another, borrowed money. Many banks use credit scoring system to evaluate their loan application. This system has a major advantage that it can sift through large quantities of credit application with minimal labor, thus reducing operating costs. A bank establishes a cutoff point which would yield the greatest net savings in loan losses. Yet credit scores provide limited information and should not be the only precursor used. Due to their small size, small banks do not have the same resources as larger banks to calculate portfolio loan loss.

The purpose of this study is to aid small community banks and credit unions in constructing a model that will predict portfolio loan risk. This study will provide valuable information to the portfolio manager of a bank which is essential to making the correct decision regarding consumer loans. We found that credit score is able to accurately identify default loans by 85%. However, they misclassify loans that defaulted as pay-off loans by almost 15%. Thus, using credit score as a sole predictor of default can be costly. Credit score can only explain 43% of probability of default. A model solely based on credit score has a low r-square and thus other variables need to be included.

Literature Review

Tufano (2009) defines consumer finance as "the study of how institutions provide goods and services to satisfy the financial functions of households, how consumers make financial decisions, and how government action affects the provision of financial services." Although in academic research corporate finance overshadows consumer finance in asset value the consumer sector dominates the corporate sector. The recent economic crisis attests to the importance of the consumer sector. Though there may be several different factors that contributed to the economic recession, without a doubt the subprime mortgage market played a big role. Thus it is important to understand consumer finance. Banks play a large role in consumer lending yet one of their greatest challenges is to find ways to prevent loan losses.

Regulations and risk management procedures are another important reason for probability of default models. One key component that bank managers are concerned with is asset management. Managers are tasked to minimize risk by diversifying their portfolio and acquiring assets with low default risk (Lopez, 2002). A big part of this is managing credit risk.

The main objective of Basel II was to revise the previous framework to be more risk-sensitive (BIS Material, 2006). They developed a three-pillar concept: 1. Minimum capital requirements, 2. Supervisory review, and 3. Market discipline. While the previous provision focused on credit risk, this new agreement also ensures that operational risk and market risk be quantified along with credit risk in order to have the appropriate capital adequacy in banks. This new accord thus attempts to strengthen the three pillars by improving risk management, governance and disclosure. Other regulation framework has been put into place in order to control bank's risk taking. After the Financial Crisis, stress testing programs have been put in place in order to test bank's ability to react to stressful situations, such as an economic crisis (BIS Material, 2011). The main objective is to put banks in a hypothetical hostile condition in order to ascertain up to what point the bank will be able to remain afloat. While stress test looks at ten different factors, one factor is the bank's exposure to default. Banks may use their own risk management default model in order to find their exposure to default. Thus it is becoming increasingly important that a bank create their own model to predict default.

Since bank loan information is hard to come by, probability of default is generally modeled using corporate securities, specifically bonds. There are two principal models in literature regarding corporate default: structural approach pioneered by Merton (1974) and reduced-form approach developed by Jarrow & Turnbull (1995). These approaches have been studied extensively and has been extended by many authors including Geske (1977), Smith & Warner (1979), Black & Cox (1976), Longstaff & Schwartz (1995), Jarrow, Lando, & Turnbull (1997), Lando (1998) and Zhou (2001). Jarrow (2009) writes a comprehensive paper comparing the structural and the reduced form models and concludes that the reduced form model is the better credit risk model. Another area of research that is related to calculating the probability of default is to valuating the recovery rate in the event that default occurs.

However, bonds and bank loans are different. While loans are monitored by bank managers, bonds must be monitored by the public who holds them. There is a dichotomy of information between the two. Where the bank has superior resources and information on their borrower, bondholder usually does not. Yet the recent growth in loan securitization has jeopardize the monitoring advantages.

Jimenez, & Saurina (2004) analyzed the determinant of probability of default focusing on three variables: collateral, type of lender, and bank-borrower relationship. They find that collateral for a loan actually increases the probability of default of a loan. While this may sound counterintuitive, their theory relates that banks tend to screen less on a loan in which collateral is provided. The risk of default was found to be affected by the type of lender, or the type of bank giving the loan. The model revealed that savings banks have a higher risk compared to commercial banks' loans. One explanation for this is that the savings' banks are controlled by managers, as opposed to commercial banks which are controlled by shareholders. Regarding relationship banking, it was found that the closer the relationship between the firm and the bank, the higher the risk of default. If a firm is being financed by only one bank and thus shows a greater commitment to that bank, the bank will be more likely to take on the risk, and thus the probability of default is higher. A thorough screening process should be employed when making loan decisions in order to avoid high default rates. Jacobson, & Roszbach (2003) create a model that determines a bank's decision on whether or not to approve a loan and the borrower's risk of default on that loan. The researchers used loan applications collected found that the income level of the applicant, whether the applicant owns a house, whether the applicant has, taxable income from a business, loans outstanding, and the existence of a guarantor all have a positive effect on whether or not the applicant gets approved for a loan. The income variable stood out because even though an applicant with a higher income was more likely to have his or her loan approved, the applicant was more likely to default on the loan. Several of the variables that affected the loan approval decision do not affect the loan's risk of default. They also found that the size of the loan itself does not alter the loan's chance of being defaulted on. A borrower was no more likely to default on a larger loan than default on a small loan. Three ratios was identified to be significant in explaining the probability of default for the data set loans: repayment capacity, owner equity, and working capital (Featherstone & Roessler, & Barry, 2006). Another discovery that was made was that as the loans aged, the probability of default decreases. This is a logical result since the longer a loan continues without defaulting, the more stable the payments have been for a longer period of time, and the default rate will be lower. Wheelock and Wilson (2000) found that banks are more likely to fail if they have low capitalization, higher ratios of loan to asset, and poor quality of loan portfolios.

The importance of consumer lending became obvious during the Great Recession. One key factor of the Great Recession was due to the increase subprime and near-prime lending, which was further aggravated by the securitization of these loans. The Financial Crisis of 2008 followed similar trends to other crisis (Demyanyk & Van Hemert, 2011). First, there was an evident boom in the subprime mortgage market. Second, a bust occurred in 2007 which is signaled by house foreclosures, high delinquencies and default rates. The subprime crisis led to spill over into other credit markets. The crisis intensified when underwriting standards deteriorated along with loan quality which led to an increase in loan risk that was not reflected by an increase in price, which led to a collapse in the market.

Thus the crisis began due to growth of subprime lending. Kwan (2001) determines that the average annual growth rate of subprime mortgages was 26 percent. Kwan concludes that subprime loans can affect credit values and the loans that are tied in with them. With an increase in subprime lending in the 2000s due to predatory practices, it was only a matter of time for a banking crisis to occur. The credit boom emanated from 2001–2006 and bust in 2007, mainly due to the large subprime securitized mortgage market (Demyanyk & Hasan, 2010). In 2008, the subprime securitized mortgage market was roughly around $1.8 trillion which is about one-third of the securitized market and 16% of total mortgage debt. Though many people doubted that such a comparatively small market could induce such a crisis, the complexity, however, of the innovated security contributed to the collapse. Keys et al. (2008) studied the link between securitization and screening subprime mortgage backed securities. They found that lenders that are most likely to securitize portfolios have less motivation to screen borrowers and more likely to default (by 10%–25%) compared to those portfolios with similar risk but with less probability of securitization. Furthermore, Mian & Sufi (2009) revealed a positive relationship between securitization and subprime lending and subsequent defaults. In geographical zones where borrowers were once denied credit, an exceptional growth of credit was offered followed by a default on those loans. In congruent with the growth in mortgage credit in this area and decrease in income growth, there was an increase to securitize these subprime mortgages. In 2008, Bernanke informed the public that 10% of near-prime mortgages and over 20% of subprime mortgages were delinquent and 2.25 million foreclosures were initiated. In 2009, these figures increased to 13% and 25% respectively. While many maligned the nontraditional features involved in mortgage contracts, Mayer, Pence and Sherlund (2009), found that the biggest reason delinquency rates were remarkably unmanageable was because it was originated to borrowers with low credit score and high loan-to-value ratios. LaCour-Little & Zhang (2014) looked at estimating the probability of default and loss given default for home equity loans around the time of the financial crisis. In this paper, they compiled data from large commercial banks, where loans were originated during 2004–2008 and tracked from 2008–2012. They are particularly interested in the relationship between loan outcomes and the lender decision to securitize the asset. After they examined loan performances, including loss given default (LGD) for home equity loans they ascertained that there was an increase in the probability of default among the particular loans that were securitized. Lending to the corporate sector through loan syndication also suffered during the 2008 Financial Crisis (Ivashina & Scharfstein, 2010). There was a 37% drop in lending during September through November period prior to the past three month period and 68% decline since the peak in 2007. The authors expostulate, however, that a decrease in lending does not necessary mean a reduction in credit supply. A decrease in lending is due to a reflection of the increase in risk. However, they noted that banks with a "strong base of deposits" will cut their lending less. Thus a bank with a solid deposit intake are inherently less risky and are capable of lending even through the crisis.

The primary risk that financial institutions face is credit risk, and thus it is essential they perform some-type of risk-taking systems (Featherstone, Roessler, and Barry, 2006). These ratings serve multiple purposes, including contributing to the loan origination process, aiding in monitoring the safety and soundness of loan portfolios, and in management reporting, facilitating adequacy of loan reserves, and providing components of loan pricing profitability analysis systems. Before the emergence of credit score, credit worthiness was measured in various ways, but normally boiled down to a judgment call (Fensterstock, 2003). A loan officer would base their decision off a system that captures the borrowers Character, Capacity, Capital, and Condition; also known as the four C's. Saunders & Allen (2002) includes another C, Collateral. Other than these factors, managers also had to take interest rate into account. However, due to the subjective nature of loan decisions, individual or business credit worthiness could vary drastically depending on the loan officer. This method looks at the customer's payment history, credit agency ratings, and financial statements among other factors (Fensterstock, 2005). However, it is inefficient because it takes copiousness amount of manual work, especially in the initial set-up of the system. Also, the weights used can be biased because of irrelevant factors that should not weigh into the situation. It is difficult of the judgmental system to determine where errors are originating from, which makes it difficult to update and correct the system.

Now, individual's risk assessment is usually given by their credit score, which is calculated by credit bureaus, Fair Isaac Corporation (FICO) being the most common. These credit scores are developed using predictive algorithms that use personal information to estimate an individuals' risk (Citron & Pasquale, 2014). FICO was created in 1956, and developed a three-digit credit score system which scores ranged from 300 to 850 where; the lower the score, the more likelihood the individual would default. In many instances, credit scores are used to price loans in order to remain objective. According to FICO, their scores are calculated using credit data which are grouped into five different categories: 1) Payment history; 2) Amounts owed; 3) Length of credit history; 4) New credit; and 5) Types of credit used.

While there are many ways to estimate credit scores, credit agency does not divulge estimation of the credit score and thus considered a "black box." Citron & Pasquale thus names three problems with the credit score system: 1) their opacity, 2) their arbitrary results, and 3) the disparate impact. Credit bureaus lack of transparency on their scoring methodology leaves individuals powerless to understand or challenge their score. Due to this opacity there exist arbitrary results. Different credit bureaus may present totally differing scores for the same individual. The secret behind the black box does not assure us of equal opportunity scoring. In fact, the scoring results show there is a disparate impact where women and minorities are concerned. Since credit score estimation is based on credit history alone, they fail to classify an entire group who may not have any credit history due to recent entrance into market, lack of large consumption in need of credit, or the fact that they rent instead of own assets. While credit score is a step in the right direction, the past financial crisis has shown that credit can still be given out incorrectly.

Methodology

Two common used methodology when working with probability of default is discriminant analysis and logistic regression. Linear discriminant analysis assumes normal distribution in the explanatory variables. Logistic regression, however, does distribution assumptions of the independent variables, thus it is more general.

Discriminant Analysis

The main objective of discriminant analysis is "to classify objects into one of two or more groups based on a set of features that describe the objects" (Gurny & Gurny 2013). In this case, we seek to classify good borrowers and bad borrowers based on different variables that describe that person. Thus the basic idea is to determine whether these different groups vary in means and if they can be used to predict default. Its primary use it to classify and make predictions where the dependent variable is in qualitative form and then find a linear combination which "best discriminates between the groups" (Altman, 1968). A disadvantage using discriminant analysis is their list of assumptions. Data is assumed to be normally distributed, variance and covariance are homogeneous, there is no correlation between means and variances, multicollinearity, and random sample. One advantage of using this method is that it reduces the space dimensionality by G - 1, where G is the number of groups. In our paper, we only have two groups (Good or Bad) and thus we have one dimension.

Discriminant analysis follows two basic steps. The first step is to estimate the coefficient of the independent variables, the borrower's characteristics. The coefficients serve as weights that measures which variables are good predictors for default. The second step is to apply a discriminant function to establish a cut-off value. The discriminant function is derived using the following equation:

$$Z = v_1 x_1 + v_2 x_2 + \cdots + v_n x_n \tag{1}$$

Where v is the discriminant coefficients and x is the independent variables. The discriminant function is treated as a standardized variable, so it has a mean of zero and a standard deviation of one. The discriminant coefficient maximizes the distances between the means of the dependent variables, where good predictor variables have the larger coefficient. Thus the discriminant function coefficient range between values of -1.0 and 1.0 and treated as a standardized variable. Thus the magnitude of the coefficients indicates the contribution of the independent variable.

An individual's z-score can be found by simply summing the product of the coefficient with the independent variable. The group mean is the average of all the individual's score, also referred to as the centroid. The success of the function can be determined by measuring the group centroid distance from one another. The key to evaluating the function is by measuring the overlap of the distribution. When the overlap of the distribution function is small the discriminant function does satisfactory in distinguishing between groups. When the overlap is large, the function has a high probability of misclassifying borrowers.

Logistic Regression

Since our objective is to find whether loan default will occur or not, than the appropriate methodology to apply would be logistic regression. Thus logistic regression takes a binary variable which only takes two values, zero or one. The main objective of a logistic regression is to find the best fitting model to describe the relationship between the dichotomous characteristic of interest (the dependent variable) and a set of independent variables. Logistic regression generates the coefficients of a formula to predict a logit transformation of the probability of the presence of the loan characteristics. A logit function thus stipulates the probability that default will occur and one minus this function specifies that default will not occur.

$$\begin{aligned} \text{Score}_i &= \alpha + b_1 x_{i1} + b_2 x_{i2} + \ldots + b_K x_{iK} \\ z &= \alpha + \sum \beta_n X_n \end{aligned} \tag{2}$$

This is a standard scoring model in which α is a constant and Xs are independent variables such as credit score, age, and other loan characteristics. In this paper since we seek to determine how credit score can find the probability of default of bank loans so our first equation will be to test this theory. Credit score ratings can be ranked in different groups taking into account 3734 the approved loan amount.

$$f(z) = \frac{1}{1+e^{-z}} = \frac{1}{1+e^{-(\alpha+\Sigma \beta_n X_n)}} \tag{3}$$

The output of this equation, which should be between 1 and 0 reveals the riskiness of the bank. An output of 0 or close to zero means the bank has low risk while an output of 1 or close to one means the bank has high risk. This logistic function can be rewritten as a logistic model by using the expression of the probability of X. Logistic regression models the probability associated with each level of the response variable by finding a linear relationship between predictor variables and a link function of these probabilities. First we need to link it with our scores variables in which

$$\text{Prob}(Default_i) = F(Score_i) \tag{4}$$

The logistic distribution function can then be written as

$$\text{Prob}(Default_i) = \frac{\exp(b'x_i)}{1+\exp(b'x_i)} = \frac{1}{1+\exp(-b'x_i)} \tag{5}$$

A very common way to estimate the weights of the coefficients is to use the maximum likelihood method. Maximum likelihood estimation is used and is the product of the sum of the logit function when default occurs multiplied by the product of the sum of one minus the logit function when the default does not happen. Then maximize the log of the likelihood function in order to find the weights:

$$(Y_i = 1) \rightarrow \text{Prob (Default}_i) = \Lambda(b'x_i)$$
$$(Y_i = 0) \rightarrow \text{Prob (No Default}_i) = 1 - \Lambda(b'x_i)$$
$$L_i = (\Lambda(b'x_i))^{y_i} (1 - \Lambda(b'x_i))^{1-y_i} \tag{6}$$
$$\ln L_i = \sum_{(i=1)}^{N} y_i \, \ln(\Lambda(b'x_i)) + (1-y_i)\ln(1 - \Lambda(b'x_i))$$

The logit model uses the logistic distribution function to link the variables. Two steps are required in order to find the coefficients: 1. Set first derivative to 0 and 2. Use the Newton's method.

$$\ln L_i = \sum_{(i=1)}^{N} y_i \, \ln(\Lambda(b'x_i)) + (1-y_i)\ln(1 - \Lambda(b'x_i))$$

$$1. \quad \frac{\partial \ln L}{\partial b} = \sum_{(i=1)}^{N} (y_i - \Lambda(b'x_i))x_i$$

$$2. \quad \frac{\partial^2 \ln L}{\partial b \partial b'} = -\sum_{(i=1)}^{N} \Lambda(b'x_i)(1 - \Lambda(b'x_i))x_i x_i' \tag{7}$$

$$b_1 = b_0 - \left[\frac{\partial^2 \ln L}{\partial b_0 \partial b_0'} \right]^{-1} \frac{\partial \ln L}{\partial b_0}$$

Lawrence & Arshadi (1995), Campbell & Dietrich (1983), Gardner & Mills (1989) all use logit models to analyze loans, in fact, Charitou, Neophytou and Charalambous (2004) states that the logit method is superior when predicting defaults.

Data

The data that will be used is from a local credit union from 2006 to December 2014. I will be looking at two different datasets: 1) current loan portfolio and 2) charge-off loans. Table 1 depicts a comparison of the loan portfolio and the charge-off loans. As of December 2014, the loan portfolio was valued $297,466,374 while the charge off loans were at $147,850.07. So roughly .05% of their portfolio loans were charged off. Table 9.1 shows the descriptive statistic of the dataset used. More information has been collected on the active loans compared to charge-off loans, we received 42,650 active loans. After cleaning up the data, there are 22,446 active loans. Information about the interest rate, original balance, current balance, loan maturity, the borrower's credit score, available credit, and loan description is given in this data set.

The charge-off loan database is a much smaller dataset with 3,371 observations. After cleaning up the charge-off data, we were left with 578 observations which can also be seen in Table 9.1. The dataset also provides information about the borrower's age and credit score, and the loan amount, duration, and loan type. We deleted observations that had missing data for credit scores and debt ratio. Also some observations seemed to be mistyped (for example an individual had a 2006 credit score value) and those were also deleted. While having more variables

Table 9.1 Descriptive statistics for active loan portfolio

Variable	Obs	Mean	Std. Dev	Min	Max
Interest Rate	22,446	8.42	5.22	0	24
Credit Score	22,446	609.18	241.22	0	964
Available Credit	22,446	87,136.25	364,944.2	0	163.36*
Original Balance	22,446	17,558.55	36,581.64	0	1.835*
Current Balance	22,446	12,873.98	46,161.29	0	4.76*
Maturity	22,446	4.38	6.21	0	36

Variable	Obs	Mean	Std. Dev	Min	Max
Credit Score	578	601.2	63.39	439	839
Age	577	41.39	12.68	21	83
Amount	578	3,536.82	4,065.38	3.2	29,443.04
Debt Ratio	578	27.38	12.86	1.37	100
Delinquency	573	463.54	247.65	2	1,002
Duration	571	2.08	1.68	0.01	13.67

Two different types of datasets were collected from a local credit union from 2006 to 2014. The top part of the table shows the summary statistics of the active loan portfolio. Current and available balances were updated December 2014. The bottom part of the table is the summary of the charge-off loans.

* Available credit, original balance, and current balance are in millions.

would be optimal, this is a good starting point and since very few researchers have the availability of bank data this will provide great insight.

Table 9.2 indicate different characteristic of good borrowers and their loans. Good borrowers are defined as borrowers who have a ranking of at least a B. This ranking defines the original risk of the loan and is assigned by the bank manager. They are assigned a binary variable of 1, and thus all who have a binary of 0 are classified as bad borrowers. Not surprisingly, the mean of a good borrower has a credit score of 647 and that of a bad borrower is 498. Thus good borrowers have significantly higher credit score. Credit score is used to price interest rate for the borrowers. Thus it makes sense that bad borrowers, who tend to have lower credit scores, also has higher interest rates. Looking at original balances granted to borrowers, the table indicate that good borrowers have a significant higher loan amount, almost a $10,000 difference, then bad borrower. This is a reasonable deduction since banks doubt bad borrower capability to pay off a big loan while trusting good borrowers' ability to take on a bigger loan and

Table 9.2 Borrower and loan characteristics

Variable	Bad Borrower [0]		Difference[0] − [1]	T-statistics
Credit Score	498.33		−149.33	[−43.28]***
Interest Rate	14.22		7.82	[103.99]***
Original Balance	10,326.65		−9,742.94	[−22.22]***
Current Balance	7,829.79		−6,795.63	[−7.44]***
Available Credit	21,604.09		−88,286.11	[−23.39]***
Maturity	3.79		−0.80	[−9.43]***

Variable	Bad Borrower [0]	Good Borrower [1]	Difference[0] − [1]	T-statistics
Auto Loan	0.416	0.451	−0.035	[−4.58]***
Personal Loan	0.249	0.122	0.127	[20.39]***
Share Secured Loan	0.024	0.012	0.012	[5.32]***
Credit Card	0.204	0.280	−0.076	[−11.99]***
End Line of Credit	0.049	0.040	0.009	[2.85]***
Home Equity Line	0.007	0.009	−0.002	[−1.46]*
Trailer Loan	0.001	0.009	−0.008	[−9.66]***
Second Mortgage Loan	0.007	0.013	−0.006	[−3.94]***
First Mortgage Loan	0.009	0.041	−0.031	[−15.75]***
Business Loan	0.009	0.000	0.009	[7.17]***
Land Loan	0.003	0.004	−0.001	[−1.31]*

Means of the borrower's loan characteristics and their significance. Using dummy variables to account the different type of loans given to borrowers, where "Good Borrowers" were those borrowers classified as B risk or above, and the rest were classified as "Bad Borrowers." Their means and differences are recorded in this table.

*** 1% statistically significant.
** 5% statistically significant.
* 10% statistically significant.

pay it off. Available credit is the difference between the credit limit of a credit card and the amount already used. Thus available credit is the share of the line of credit that has not been spent. As for the case of original balance, the available credit is statistically significantly larger for good borrowers than bad borrowers. Good borrowers thus are granted a higher limit than bad borrowers. Maturity of the loan in this study can be defined as the length of the life of the loan. This was found by looking at the original date and the due date of the loan. Thus this figure shows that good borrower tend to have loans with longer maturity than bad loans. In order to reduce the risk of bad borrower loans they will give them a loan with shorter maturity.

This table also presents the different types of loans that this particular credit union gives to consumers. A binary variable was created to indicate what type of loans borrowers took out. The loans available are: auto, personal, share, credit card, end line credit, home equity line, trailer, second mortgaged, first mortgage, land, or business loan. Another possible explanation of the larger loan amount for good borrowers is the fact that good borrowers tend to invest in more expensive tangible items such as autos, homes, and land. Table 9.4 shows that good borrower take out more auto loans, home equity lines, trailers, first and second mortgages, and more land loans. In addition to these investments, good borrowers have more credit cards. Bad borrowers, on the other hand, tend to take out more personal loans, share secured loan, open end line of credit and business loans compared to good borrowers.

Table 9.3 compares the means of the variables for the charge-off dataset categorized into loan type. For the age characteristic, it is not significant for any loan type except for achiever loans. The mean age for most charge-off borrowers are in the lower 40s. Therefore, the idea that younger borrowers are more likely to default on their payment is not substantiated by looking at just the means. The achiever loan is the only loan that is statistically significant and that its age is lower than 40s. Achiever loan borrowers are seeking to build credit and one type of borrower who lacks most in credit history are young adults who have not had the opportunity to build a history. The charge-off amount is the statistically significant for all the loan types. The highest charge-off amounts are from auto loans (new, used and indirect loans). Since borrowers take out larger loans to afford an auto, it stand to reason that they will have higher charge-off amounts. The achiever loan is the lowest charge-off amount. Other than the fact that there is only few observations, achiever loans by definition is a small loan with the sole purpose to build up credit. The credit score variable is significant for all loan types except for used and new auto loans. Indirect auto and credit card have the highest credit score. Indirect auto loans usually are originated in the car dealership and then transferred to the bank. Thus the bank does not have direct contact with the buyer. Credit card application, while many times dealt through the bank, also has a third party involved, the credit card company (in this case Visa). Thus these third party loan transactions may require a larger credit score cut-off before being accepted. Personal loans have a lower credit score than the other loans. Relationship banking may have influenced the acceptance of this

Table 9.3 Charge-off loan characteristics

Variable	Other Loans [0]	Personal Credits [1]	Difference[0] − [1]	T-statistics
Age	40.74	42.34	−1.60	[−1.461]
Amount	4,716.82	1,630.66	3,086.16	[11.369]***
Credit Score	608.56	589.30	19.26	[3.621]***
Debt Ratio	26.58	28.67	−2.09	[−1.827]*
Delinquency	409.25	551.95	−142.70	[−6.886]***
Duration	2.65	1.14	1.51	[13.648]***

Variable	Other Loans [0]	Credit Cards [1]	Difference[0] − [1]	T-statistics
Age	41.24	42.51	−1.27	[−0.688]
Amount	3,642.60	2,443.78	1,198.82	[3.04]***
Credit Score	599.68	616.90	−17.22	[−1.943]**
Debt Ratio	27.28	28.42	−1.14	[−0.651]
Delinquency	452.33	583.41	−131.08	[−3.796]***
Duration	1.96	3.36	−1.40	[−2.93]***

Variable	Other Loans [0]	Indirect Auto [1]	Difference[0] − [1]	T-statistics
Age	41.59	40.03	1.56	[0.989]
Amount	3,253.52	5,135.66	−1,882.14	[−4.167]***
Credit Score	598.29	617.60	−19.31	[−2.851]***
Debt Ratio	27.78	25.13	2.65	[2.148]**
Delinquency	489.06	321.00	168.06	[6.923]***
Duration	2.04	2.30	−0.26	[−1.665]*

Variable	Other Loans [0]	Used Auto [1]	Difference[0] − [1]	T-statistics
Age	41.86	40.19	1.67	[1.491]
Amount	2,811.39	5,207.39	−2,396.00	[−5.817]***
Credit Score	600.14	603.62	−3.48	[−0.598]
Debt Ratio	27.98	25.99	1.99	[1.920]**
Delinquency	487.95	408.02	79.93	[3.749]***
Duration	1.87	2.57	−0.70	[−5.188]***

Variable	Other Loans [0]	New Auto [1]	Difference[0] − [1]	T-statistics
Age	41.28	43.24	−1.96	[−0.854]
Amount	3,396.43	7,260.43	−3,864.00	[−2.672]***
Credit Score	601.00	606.38	−5.38	[−0.473]
Debt Ratio	27.43	25.98	1.45	[0.476]
Delinquency	468.66	328.90	139.76	[3.054]***
Duration	2.03	3.41	−1.38	[−3.932]***

Variable	Other Loans [0]	Achiever [1]	Difference[0] − [1]	T-statistics
Age	41.47	31.71	9.76	[2.842]**
Amount	3,579.81	30.45	3,549.36	[20.816]***
Credit Score	602.39	503.86	98.53	[5.672]***
Debt Ratio	27.14	46.63	−19.49	[−1.542]
Delinquency	466.97	186.29	280.68	[5.443]***
Duration	2.10	0.35	1.75	[18.405]***

Dummy variables are assigned to different loan types in the charge-off dataset. The means of variables are then compared by specific loan versus the rest of the loans.

*** 1% statistically significant.
** 5% statistically significant.
* 10% statistically significant.

loan application. An achiever loan has the lowest credit score. Credit cards and personal loans have higher payment delinquency than the other loan types. These loans, which also have lower charge-off payment (and loan amount), are given more time in delinquency until marked off the books. Duration is the amount of time that the borrowers paid off their loan before defaulting. The three auto loan and credit cards loans have higher duration mean the other loan type. Thus they had more loan payment periods than the other loans.

Using these two dataset, we will attempt to see how and if credit score is a good measure for default. While some charge-offs are due to the borrower's death or incarceration, most are due to bankruptcy, post repo, inability to pay, or just the refusal to pay. In order to run a test to see whether credit score can really predict the probability of default we need a database that has both default loans and paid-off loans. In our portfolio loan database we find which loans have paid of at least 99% of their loans back and we assign the binary variable of 0. We merge them with our defaulted loans, classified as 1, and delete any default that was due to death, prison, or any charge-off amount below $50 or have been delinquent in the last 30 days. Our new database has 1261 observations with 543 being default observations and 718 being paid-off loans. The credit score ranges from 437 to 850 with a mean of 677. The only shared variable is credit score and thus our regression will focus on the influence that credit score has as a predictor of probability of default.

Results

While we have run analysis on good borrower, we are more interested in how to identify a borrower with default potential. We use our merged data to get a clearer picture of whether credit score can predict the probability of default which is shown in Table 9.4. Since credit score is our only variable in this database we are constricted in using just this predictor. The function again is significant and because only one predictor is use, the standardized canonical coefficient and structure is one. The unstandardized canonical coefficient is given in which we are able to find the discriminant score function: $D_i = -9.99 + 0.015 \times X_i$, where i is each individual borrower and x is their credit score. The idea behind the discriminant score is to find the group means. The group means for paid-off loans is 0.835 and for default loans, -1.104. Credit score is able to correctly classify loans that will default (be paid-off) by 85% (81%). Thus this model is relatively good. A manager has a reasonable vindication to assign a cut-off score to avoid charge-offs. Thus a manager will be willing to accept a credit score which discriminant score is closer to group mean of 0.835. The group mean's credit score is approximately a credit score of 722, thus anything above that should clearly be accepted. The average credit score of this database is a 677, even though the descriminant score is below the group mean, the discriminant score of .165 is much closer to the group zero's mean than group one's mean. Thus a credit score of 677 should also be accepted.

Table 9.4 Probability of default using discriminant analysis

Fcn	Canon. Corr.	Likelihood Ratio	F	Prob>F
1	0.693	0.52	1162.8	0.000
Ho:	This and smaller canon. Corr. Are zero			
		Unstandardized Canonical	Standardized Canonical	Canonical structure
	Credit score	0.015	1	1
	constant	−9.99		
	Default Loans	Group Means		
	0	0.835		
	1	−1.104		
	TRUE	Classified		
	Default Loans	0	1	Total
	0	580	138	718
		80.78%	19.22%	100%
	1	79	464	543
		14.55%	85.45%	100%

Discriminant analysis is run on a merged dataset of defaulted loans and paid-off loans. The defaulted loans were given the binary value of 0 and the paid-off loans were given the binary number of 1. There is only one predictor value: credit score. The first section is the canonical test, which shows the number of functions and its significance. The second section looks at the unstandardized and standardized canonical coefficient for the discriminant function and the canonical structure. The next section looks at the group means: the "centroids." The last section looks at the classification, or whether the model is able to classify correctly the type of borrower.

Though credit score does an admirable job at classifying loans probability of default, there is still room for improvement. It still has a 19% probability that it will classify a loan as a pay-off loan when it actually will default. Depending on the amount of charge-off, this can be a huge loss for any bank. In order to make this model more accurate, more variables will need to be added in order for the model to discriminate more efficiently.

$$\text{prob}(\text{Default}) = \alpha + \beta_1 \text{Credit Score}$$

We also run a logistic regression run is on the merge data in order to find the likelihood of a borrower defaulting. The results can be found in table 9.5. Credit score variable is negatively significant with a coefficient of -0.024. Thus a one unit increase in credit score, will produce a .024 decrease likelihood that the loan will default. Looking at the odds ratio, a one unit increase in credit score is associated with a 2.4% odd of decrease probability of default. If we wanted to find the probability of the individual defaulting we can plug the coefficients in to the equation: $Y = \ln\left(\dfrac{p}{1-p}\right) = 15.923 + -.024 * (\text{credit score})$. If we use the mean of our database, which is x = 667, than Y = -.085. Since we are looking for the

Table 9.5 Logistic regression of probability of default

Prob	Coef	Std. Err.	P-Value
Credit Score	−0.024	0.001	0.001
Constant	15.923	0.852	0.001
N	1,261		
Pseudo R²	43.37%		

Prob	Odds Ratio	Std. Err.	P-Value
Credit Score	0.976	0.001	0.001
Constant	8,253,401	0.852	0.001

Logistic regression on merge data set of default and paid-off loans. The default loans were given a binary value of 1 while the paid-off loans were given the value of 0. The probability of default is the dependent variable and credit score is the sole independent variable.

Table 9.6 Result summary for probability of default

Credit Score	Discriminant Score	Probability of Default
500	−2.490	98.1%
592 *(DS mean)*	−1.110	84.7%
663	−0.045	50.3%
667	0.015	47.9%
722 *(DS mean)*	0.840	19.7%
735	1.035	15.2%
786	1.800	5.0%
850	2.760	1.1%

probability we will have to transform that product as well: $p = \left(\dfrac{e^y}{e^y + 1}\right) = 41.9\%$

that a borrower with a credit score of 667 will default. Thus the probability of default of the average credit score borrower in this dataset (creditscore = 677) is 41.9%. Obviously, the higher the credit score the less probability of default. An 850 credit score borrower would have a 1.1% probability of default while a borrower with a 500 credit score would have a 98.1% chance of default. However, the Pseudo R² is only 43% so other variable should be taken into account.

The discriminant score mean had a score of 722, which has a 19.7% probability of default. If a bank want to lower their probability of default they must choose a score that is close to discriminant score mean, since this function does a good job of classification. The greater the credit score the lower the probability of default as can be seen in Table 9.6. This function and probability of default model however can be improved by adding other variables. Thus a bank can decide on the risk they are willing to take with borrowers in order to identify a credit score hurdle.

Conclusion

Most of the literature related to probability of default has focused on the bond market. The consumer lending market has been gravely overlooked in research mainly due to data available. Consumer lending however is a huge market that should not be overlooked. Due to the recent financial crisis banks have suffered a massive loss on loans. This brings up an awareness of a need for an approach that banks can adopt to mitigate these losses and to measure credit risk more accurately. The objective of this paper is to find a simple model that small bankcs could use in order to find the probaiblity of default of their loans.

Credit score is a wonderful tool to measure riskiness of a borrower. Using discriminant analysis and logistic regression we found that credit score is a predictor of default. Using discriminant analysis a manager could assign a cut-off score that will reduce the likelihood of default. However, its opaqueness and the fact the borrowers with good score default shows that it should not be the sole predictor. While the credit score model had passable result in classifying loans, they still misclassified loans. This error could cause a bank to lose money. The logistic regression also had a low r-square which also causes question to the accuracy of the model. In order to have a better predicting model, more variable should be included.

Adding more variables to the model will help in the accuracy of the prediction. This gives practical use to bankers when making the decision of whether to accept a loan. While credit score does give good information, it should not be the sole factor when making this decision. This model can also be used for existing loan. If a bank knows the probability of default of their existing loan they will be forwarned therefore forearmed in case the worse were to occur. Having this information is essential for a bank to make the proper credit rationing and capital adequacy decisions.

References

Altman, E. I. (1968). Financial ratios, discriminant analysis and the prediction of corporate bankruptcy. *The journal of finance, 23*(4), 589–609.

Altman, E. I., Gande, A. and Saunders, A. (2010). Bank debt versus bond debt: Evidence from secondary market prices. *Journal of Money, Credit and Banking, 42*(4), 755–767.

BIS Material (June 2006). Basel II: International convergence of capital measurement and capital standards: A revised framework – Comprehensive version. *Bank for International Settlement.*

BIS Material (June 2011). Basel III: A global regulatory framework for more resilient banks and banking systems – Revised Version June 2011. *Bank for International Settlement.*

Black, F., & Cox, J. C. (1976). Valuing corporate securities: Some effects of bond indenture provisions. *The Journal of Finance, 31*(2), 351–367.

Campbell, T. S., & Dietrich, J. K. (1983). The determinants of default on insured conventional residential mortgage loans. *The Journal of Finance, 38*(5), 1569–1581.

Charitou, A., Neophytou, E., & Charalambous, C. (2004). Predicting corporate failure: empirical evidence for the UK. *European Accounting Review, 13*(3), 465–497.

Citron, D. K., & Pasquale, F.A. (2014). The scored society: Due process for automated predictions. *Washington Law Review, 89.*

Demyanyk, Y., & Hasan, I. (2010). Financial crises and bank failures: A review of prediction methods. *Omega, 38*(5), 315–324.

Demyanyk, Y., & Van Hemert, O. (2011). Understanding the subprime mortgage crisis. *Review of Financial Studies, 24*(6), 1848–1880.

Fensterstock, A. (2003). Credit scoring basics. *Business Credit* Mar. 1 2003. pp. 47–78.

Featherstone, A.M., Roessler, L. M., & Barry, P.J. (2006). Determining the probability of default and risk-rating class for loans in the seventh farm credit district portfolio. *Applied Economic Perspectives and Policy, 28*(1), 4–23.

Gardner, M. J., & Mills, D.L. (1989). Evaluating the likelihood of default on delinquent loans. *Financial Management,* 55–63.

Geske, R. (1977). The valuation of corporate liabilities as compound options. *Journal of Financial and Quantitative Analysis, 12*(04), 541–552.

Gurný, P., & Gurný, M. (2013). Comparison of credit scoring models on probability of default estimation for US banks. *Prague Economic Papers, 2013*(2), 163–181.

Ivashina, V., & Scharfstein, D. (2010). Bank lending during the financial crisis of 2008. *Journal of Financial Economics, 97*(3), 319–338.

Jacobson, T., & Roszbach, K. (2003). Bank lending policy, credit scoring and value-at-risk. *Journal of Banking & Finance, 27*(4), 615–633.

Jarrow, R.A. (2009). Credit risk models. *Annual Review of Financial Economy, 1*(1), 37–68.

Jarrow, R.A., Lando, D., & Turnbull, S.M. (1997). A Markov model for the term structure of credit risk spreads. *Review of Financial Studies, 10*(2), 481–523.

Jarrow, R.A. and Turnbull, S.M. (1995). Pricing derivatives on financial securities subject to credit risk. *The Journal of Finance, 50*(1), 53–85.

Jiménez, G., & Saurina, J. (2004). Collateral, type of lender and relationship banking as determinants of credit risk. *Journal of Banking & Finance, 28*(9), 2191–2212.

Keys, B., Mukherjee, T., Seru, A., & Vig, V. (2008). Securitization and screening: Evidence from subprime mortgage backed securities. *Quarterly Journal of Economics, 125*(1).

Kwan, Simon. (2001). Rising Junk Bond Yields: Liquidity or Credit Concerns? *FRBSF Economic Letter* 2001-33 (November 16).

LaCour-Little, Michael, and Yanan Zhang. (2014). Default Probability and Loss Given Default for Home Equity Loans. *Economics Working Paper. SpringerReference.* Office of the Comptroller of the Currency.

Lando, D. (1998). On Cox processes and credit risky securities. *Review of Derivatives research, 2*(2–3), 99–120.

Lawrence, E. C. and Arshadi, N. (1995). A multinomial logit analysis of problem loan resolution choices in banking. *Journal of Money, Credit and Banking,* 202–216.

Longstaff, F. A., & Schwartz, E. S. (1995). A simple approach to valuing risky fixed and floating rate debt. *The Journal of Finance, 50*(3), 789–819.

Lopez, J.A. (2002). The Empirical Relationship between Average Asset Correlation, Firm Probability of Default and Asset Size. Working Paper, Federal Reserve Bank of San Francisco.

Mayer, C., Pence, K., & Sherlund, S.M. (2009). The rise in mortgage defaults. *The Journal of Economic Perspectives, 23*(1), 27–50.

Merton, R.C. (1974). On the pricing of corporate debt: The risk structure of interest rates. *The Journal of Finance, 29*(2), 449–470.

Mian, A., & Sufi, A. (2009). *The consequences of mortgage credit expansion: Evidence from the US mortgage default crisis.* The Quarterly Journal of Economics, *124*(4), 1449–1496.

Saunders, A., & Allen, L. (2002). *Credit risk measurement: new approaches to value at risk and other paradigms* (Vol. 154). John Wiley & Sons. 9–11.

Smith, C. W., & Warner, J. B. (1979). On financial contracting: An analysis of bond covenants. *Journal of Financial Economics, 7*(2), 117–161.

Tufano, P. (2009). Consumer finance. *Annual Review Finance Economy, 1*(1), 227–247.

Wheelock, D. C., & Wilson, P. W. (2000). Why do banks disappear? The determinants of US bank failures and acquisitions. Review of Economics and Statistics, *82*(1), 127–138.

Zhou, C. (2001). The term structure of credit spreads with jump risk. *Journal of Banking & Finance, 25*(11), 2015–2040.

10 Intelligent distributed applications in e-commerce and e-banking

Jennifer Brodmann and Phuvadon Wuthisatian

E-commerce

The concept of electronic commerce, also known as e-commerce, came into the business vocabulary in the 1970s (Wigand, 1997). The widespread use of personal computers in the internet age has created a paper-free environment for the common person. Wigand (1997) discusses in a broad sense that e-commerce "includes any form of economic activity conducted via electronic connections." The means in which the market is coordinated is the commonality that categorizes it as e-commerce, with the formal definition being

> the seamless application of information and communication technology from its point of origin to its endpoint along the entire value chain of business processes conducted electronically and designed to enable the accomplishment of a business goal. These processes may be partial or complete and may encompass business-to-business as well as business-to-consumer and consumer-to-business transactions.

Gunasekaran et al. (2002) state that in order for a firm to fully reap the benefits of e-commerce adoption, "firms must understand its potential, its components, their own businesses, and the businesses of trading partners," and conclude that "development of effective strategies for achieving competitive advantage through EC will be necessary for success in the 21st century." Many firms have adopted e-commerce as a means of being competitive, and there are now firms whose business models are based on e-commerce platforms. Examples include tech behemoths such as Amazon.com, fintech firms such as SoFi and Lending Club, and insurance companies such as Esurance.

The emergence of e-commerce and e-banking

E-commerce emerged from the development of the internet and the World Wide Web and has become the universal means of conducting business. The finance industry has widely adopted e-commerce as a way to conduct financial transactions, manage accounts, and conduct communication among parties. Online markets can reduce barriers to entry for new market participants. In addition, web search engines and online business directories reduce search costs for price comparisons and service quality and reputation (Bakos, 2001).

E-commerce adoption

There have been several studies that examine the firm attributes that relate to e-commerce adoption. Consumers may be more likely to purchase from the web if they are more experienced in using the web and if they perceive a higher level of trust in e-commerce (Corbitt et al., 2003).

Chatterjee et al. (2002) propose three main factors that influence an organization shaping e-commerce activities: top management championship, strategic investment rationale, and the extent of coordination, which they find helps foster assimilation at a high level. Daniel et al. (2002) find sequential stages for internet and e-commerce adoption, which are initial e-commerce service development; utilizing email to communicate with customer, suppliers, and employees; utilizing information-based websites and engaging in online ordering facility development; and having online ordering in operation and development of online payment capabilities. Daniel and Grimshaw (2002) study why small and large firms are adopting e-commerce and provide a comparison of what the benefits are that these two segments of firms are experiencing from it. They find that the main reasons behind the firms adopting commerce stem from responding to competition, enhancing customer service, and boosting supplier relationships, which was more pronounced for small firms, except for improving operational efficiency, which was more pronounced for large firms.

E-commerce diffusion for business-to-business is driven by global forces while business-to-commerce is driven by more local forces, since business-to-business is influenced by multinational corporations pushing e-commerce to their international suppliers, consumers, and foreign subsidiaries, while business-to-consumer is influenced by the consumer market, which is mostly local (Gibbs et al., 2003). For small and medium-sized enterprises (SMEs), the perceived strategic value of e-commerce may influence e-commerce adoption (Grandon and Pearson, 2004).

Eastin (2002) examines the adoption of four main e-commerce activities – online shopping, online banking, online investing, and electronic payment for access to exclusive websites – and finds six common attributes for e-commerce diffusions: "perceived convenience and financial benefits, risk, previous use of the telephone for a similar purpose, self-efficacy, and Internet use." He also finds that when users adopt one of the four e-commerce activities, they are more likely to adopt another. Bhattacherjee (2000) studies what motivates consumers to adopt business-to-consumer e-commerce services and finds that a modified theory of planned behavior (TPB) helps explain e-commerce service acceptance. TPB defines the intentions with three structures of belief: attitude, subjective norm, and behavior control.[1] Gibbs and Kraemer (2004) examine the factors that influence e-commerce adoption and find that "technology resources, perceived strategic benefits, financial resources, and external pressure" are the main determinants of adoption as well as the scope of e-commerce adoption. Hong and Zhu (2006) state that "technology integration, web functionalities, web spending, and partner usage" mainly distinguish firms from being e-commerce adopters. Egger (2001) studies a model for trust in e-commerce (MoTEC) and finds that the design interface of the website is only one of part of the entire customer experience and attention is also paid to the general management and marketing to build consumer trust in e-commerce.

E-commerce quality

Web consumers that utilize e-commerce have different attributes than brick-and-mortar consumers, and this phenomenon has been studied frequently in the recent literature. Barnes and Vidgen (2002) conduct a study of e-commerce quality using the WebQual index and identify five factors, which include usability, site design, information quality, trust, and empathy. Security and privacy are the main reasons behind dissatisfaction in using e-banking, while perceived ease of use and perceived usefulness increase customer satisfaction (Jalal et al., 2011). Consumers may have privacy concerns, which can impede e-commerce transactions, but the cumulative influence of internet trust as well as personal interest in the internet can offset the privacy risk perception in a consumer information disclosure when using the internet (Dinev and Hart, 2006). One of the necessary components of e-commerce is user authentication to prevent fraud, and because of this there is a loss of the consumer's privacy (Gunasekaran et al., 2002). Strict enforcement of a privacy policy may help resolve this. Chen and Dhillon (2003) consider the sources of trust in e-commerce to be consumer characteristics, the firm, the firm's website, and the communication between the firm and the consumer, and they propose a path model in order to understand how to have consumers instill trust in utilizing e-commerce.

E-banking

E-banking definition

Hertzum et al. (2004) define e-banking as "web-based electronic banking." This is becoming a popular means of banking, and now there are consumers that solely do their banking through "digital banks," such as Ally Bank. E-banking can allow users to conduct bank transactions through bank websites, such as sending money, check bank account balances, disputing charges, and printing bank statements, to name a few.

E-banking quality

The most important quality attributes of internet-based e-banking usefulness are "accuracy, security network speed, user-friendliness, user involvement and convenience" (Liao and Cheung, 2002). Loonam and O'Loughlin (2008) propose that future e-banking and complex financial product adoption will be contingent upon assertive online marketing campaigns and a rise in the responsiveness of e-banking websites. Bauer et al. (2005) examine the factors that determine e-banking quality and find that this comprises three dimensions: security trustworthiness and essential services, attractive cross-buying services and added value, and transaction support and service provider responsiveness. Satisfaction from previous usage of the bank website has a positive effect on consumer loyalty and positive word-of-mouth (Casaló et al., 2008). Website usability has a positive impact on customer satisfaction as well as loyalty having a positive relationship with positive word-of-mouth Casaló et al., 2008). Service quality is needed to boost customer satisfaction and to build trust and loyalty (Chu et al., 2012).

Advantages and disadvantages of e-banking

E-banking has allowed for ease of bank transactions, account management, and remote access. This remote access to banking has been seen as appealing for rural bank customers. Yet rural consumers base their bank selection on personal acquaintance compared to city consumers, because rural consumers more probably take a loan out of their bank (De Blasio, 2008). However, the disadvantages focus mainly on security and confidentiality risks. Hacking and identity theft are issues that banks must address to maintain consumer confidence.

E-banking adoption may also differ by generation, with older generations such as baby boomers preferring to utilize e-banking through computers and younger generations, such as millennials preferring to access bank services through mobile applications (Brodmann et al., 2018). Perceived difficulty using computers and lack of personal service in e-banking were the major barriers for mature customers (Mattila et al., 2003). Gefen (2000) develops a MoTEC and finds that consumer familiarity with an internet vendor may help build confidence, but it is also the consumer's disposition to trust that will influence whether a consumer will use an internet vendor.

Current trends in e-banking

There has been a drastic increase in consumers using e-banking and m-banking services as their primary means of managing their bank accounts. Close to half of adult Europeans (51%) utilize internet banking, which is up significantly from 25% in 2007.[2] This may stem from the growth in consumers using e-commerce for the majority of their business transactions because of ease of use and convenience. A large proportion of the US population has access to the internet, a personal computer, and a smartphone. The increase in mobile phone usage by many Americans and the evolution of mobile banking may allow for increases in financial inclusion for underserved bank customers (Gross et al., 2012).

The Federal Reserve report, using a 2015 survey, finds that mobile banking use is continuously rising, with 43% of all mobile phone users that have a bank account using mobile banking prior to taking the survey.[3]

Services provided from e-banking

Referencing Drigă and Isac (2014), we will briefly discuss the major types of services in e-banking.

Home banking

Home banking is any financial services that a bank customer can do at home, which includes contacting a bank representative over the phone to facilitate any bank account inquiries, customer service, resetting of account passwords, money transfers, and so forth.

Personal computer (PC) banking

PC banking is a type of service where bank customers can manage their accounts and financial services through a bank's proprietary software, which they can use on their personal computer by using a modem.

Internet banking

Internet banking is a service where bank customers can remotely access their accounts and manage them through the bank website and facilitate bank transfers, make electronic payments, check account balance, communicate with customer service, and access general information.

Mobile banking

This service allows bank customers to manage their accounts through their mobile device, which is a smartphone with access to wireless internet or cellular data. They can manage accounts, pay bills, check account balances, and make check deposits through this system.

Security issues related to e-banking

Phishing

Online service provides a very comfortable life for people, especially in banking. Most banks switched from the traditional approach, in which people conduct all transactions with physical banks, to online service or E-banking, which allows customers to conduct all transactions online in order to reduce their time to travel and get things done in the physical banks. The online services, however, do not mean to provide the safest way to protect users from online thieves, especially phishing.

Phishing refers to illegal activities launched by miscreants who are trying to make profits by means of illegal financial transactions from the users.[4] In general, phishing tricks users into providing information regarding their financial activities, such as passwords, identifications, and other confidential information. The Anti-Phishing Working Group (APWG), the organization that is focusing on eliminating the phishing threats and frauds, reports phishing attacks during the first quarter of 2018, indicating that the unique phishing sites were higher than the last year and these phishing attacks targeted mainly payment services as well as email, which accounted for more than 50% of all phishing attacks.

The US Federal Bureau of Investigation (FBI) has developed an internet crime report called "FBI's Internet Crime Complaint Center" (IC3) to receive internet crime complaints. The report in 2017 shows that IC3 received more than 300,000 complaints and the approximate loss from the victims was accumulated more than $1.42 billion. The primary target for this is businesses that work with foreign suppliers and/or perform wire transfer payments. It is essential for banking security to protect consumers' identities and all related confidential information.

Why are banks concerned about phishing attacks? Suh and Han (2003) test for the factors that severely affect banks from phishing attacks and find that the most harmful effect is "trust crisis." The trust crisis refers to the eroding process of customers' trust in the banks. Once the banks cannot maintain the trust of customers, then bank failure will occur. Most banks have developed a security system to protect customers' confidential information such as security questions, picture passwords, email confirmation, and passcodes. The goal is to protect the consumers' confidential information as well as to develop trust in the consumers' minds.

Phishing in numbers

Phishing attacks are increasing rapidly in recent years due to the access of internet users as well as the growth in E-banking. According to APWG report, it shows that there were 263,538 unique phishing attacks in the first quarter of 2018, increasing approximately 46% from the fourth quarter of 2017.

The primary targets of these attacks were in payment services and online banking. Interestingly, the email attacks were increasing from 16% to 18.7%, indicating that the access to the internet of consumers, while growing rapidly, causes the phishing attackers to target more. Payments, however, decreased during the period from 42% to 39.4%, as most banks developed security systems to protect the customers from these phishing attacks.

Types of phishing

DECEPTIVE PHISHING

Attacks that use spoofed email to get personal information, such as credit card numbers, account usernames and passwords, and social security numbers from the victims is called "deceptive phishing."[5] This type of phishing often comes in the form of a legitimate company's emails. For example, PayPal scammers may send the email to instruct the users to either click on the link or login to the fake PayPal page to get the user's information.

Ali and Rajamani (2012) conduct the experiment using the Anti-phishing Detector System (ADS) in instant messagers (IMs). They find that using ADS can detect more than 70% of phishing attacks in IMs. Their finding, however, indicates that there are many users that fall for deceptive phishing, and they are unable to differentiate this type of phishing.

SPEAR PHISHING

Spear phishing is an email-spoofing attack that targets specific individuals or organizations to get access to sensitive information.[6] By definition, it is similar to deceptive phishing. However, spear phishing is, most of the time, conducted by the preparators and it is more sophisticated than deceptive phishing. The organizations try to prevent this type of phishing since these attacks can get a company's confidential information.

PHARMING

Pharming refers to a cyber-attack that is intended to redirect a website's traffic to another site or a fake site. This method modifies the domain name system (DNS) so that users think they are accessing a legitimate website. However, in fact, the site is directing to the fraudulent one. Once users put their credential information in these fraudulent websites, the criminals will have the users' information.

FRAUD

Bank fraud is, according to US Code § 1344, the use of illegal means to obtain money, funds, credits, assets, securities, or other property owned by, or under the custody or control of, a financial institution. Bank fraud, in fact, reduces the creditability of the banks greatly, since customers value the trust in the banking system.[7]

Types of fraud

IDENTITY THIEF OR IMPERSONATION

An identity thief operates by obtaining confidential information of individuals and using the information to apply for credit cards or open multiple accounts under that person's name. The typical information that fraudsters obtain are names, dates of birth, social security numbers, and tax identification numbers. The information is sufficient enough to open accounts or credit cards. The identity thief can be either an outsider or insider (e.g. a bank employee uses the customer's information to open the new accounts for them in order to receive a higher commission).

MONEY LAUNDERING

Money laundering refers to the act of concealing the transformation of profits from any illegal activities into "legitimate" assets. The typical illegal activities include drug trafficking, terrorist activities, or other crimes that involve money from illegal businesses. The fraudsters use the illegal money to obtain economic assets such as properties, mortgages, and others that can create wealth in the mainstream economy.

ACCOUNTING FRAUD

Some financial institutions try to hide their financial problems by creating false information to overstate their value or profit, although they are operating at a loss. The financial institutions do so in order to raise capital or seek investments to cover their losses. For example, Enron – the former US energy, commodities, and services company based in Texas – provided false financial statements for its shareholders and potential investors, stating that the company significantly improved its profits and had expansion opportunities overseas. However, the company hid billions of dollars in debts and failed to deliver the promised investments both in the United States and abroad. The stock price dropped from a high of $90.75 to less than a dollar by the end of November 2001.[8]

In 2016, Wells Fargo was issued a total of $185 million in fines for creating more than 1.5 million checking and savings accounts and more than 500,000 credit cards for which customers never gave authorization. Wells Fargo employees anticipated incentive-compensation programs by opening new accounts and credit lines for customers. This allegation led to more than 5,000 employees being fired, and the bank was sued by many organizations such as the Consumer Financial Protection Bureau, the Los Angeles City Attorney, the Office of the Comptroller of the Currency, and individual customers. After the scandal, the level of trust in the bank declined dramatically, and the bank reported a loss in profits of 19% in the last quarter of 2017.[9] The effects of the scandals did trigger loss of customers' trust in banks as well as investors, and many decided to have no further related activities with the bank.

Wells Fargo and Enron are examples of how creditability of the bank can be ruined by instances of fraud. As explained by Suh and Han (2002), trust is a vital element of banks, and once the trust is in crisis, it affects the bank's performance and its future investments, and makes it difficult to gain trust back from customers.

Conclusion

The American Bankers Association[10] reported an increase in the number of online banking users and the number is expected to increase in the future due to the access of the mobile internet as well as the online banking services. E-banking is evolving and is expected to be part of banking customers. E-banking provides users a means of tracking their spending, income, and their investments in no time, simply by logging into their accounts and looking for historical activities. There is no doubt that online banking services can make life much better, and customers do not need to visit the physical banks to conduct transactions that take time that they would rather spend on something more productive.

However, online banking also comes with risks. Some people try to get credential information from banking users to gain their personal wealth or use the information to conduct illegal activities. It is imperative for banks to create security in order to protect its customers' information. At the same time, customers should learn how to protect themselves against any potential fraudsters, spam, or phishing attacks. Although the bank security system is implemented to protect the users, users are also required to ensure that their information is safe. In the meantime, banks should provide transparency to customers to prevent any fraud that could happen to customers' information.

Notes

1 Bhattacherjee (2000) describes attitude as the inclination towards a certain object, event, or act that would result in an actual behavior; subjective norm as ideas of social forces impacting a behavior; and behavioral control as ideas of internal and external constraints that influence behavior.
2 Internet banking on the rise. (2018, January 15). Retrieved September 2, 2018, from https://ec.europa.eu/eurostat/web/products-eurostat-news/-/DDN-20180115-1
3 Consumers and Mobile Financial Services Report 2016 (Federal Reserve).
4 Aburrous et al. (2010).
5 Huang, Tan, and Liu (2009).
6 Parmar (2012).
7 Suh and Han (2003) and Dyck, Morse, and Zingales (2010) report bank frauds as the factor that customers switch to other banks and that these frauds reduce the value of the bank.
8 www.nytimes.com/2007/05/10/business/worldbusiness/10iht-enron.1.5648578. html?_r=0
9 www.usatoday.com/story/money/2017/10/13/wells-fargo-earnings-report/759138001/
10 www.aba.com

References

Aburrous, M., Hossain, M.A., Dahal, K. and Thabtah, F. (2010). Experimental case studies for investigating e-banking phishing techniques and attack strategies. *Cognitive Computation*, 2(3), 242–253.

Ali, M. and Rajamani, L. (2012). APD: ARM deceptive phishing detector system phishing detection in instant messengers using data mining approach. In *Global Trends in Computing and Communication Systems* (pp. 490–502). Springer, Berlin, Heidelberg.

Bakos, Y. (2001). The emerging landscape for retail e-commerce. *Journal of Economic Perspectives*, 15(1), 69–80.

Barnes, S. J. and Vidgen, R.T. (2002). An integrative approach to the assessment of e-commerce quality. *Journal of Electronic Commerce Research*, 3(3), 114–127.

Bauer, H. H., Hammerschmidt, M. and Falk, T. (2005). Measuring the quality of e-banking portals. *International Journal of Bank Marketing*, 23(2), 153–175.

Bhattacherjee, A. (2000). Acceptance of e-commerce services: the case of electronic brokerages. *IEEE Transactions on Systems, Man, and Cybernetics-Part A: Systems and Humans, 30*(4), 411–420.

Brodmann, J., Rayfield, B., Hassan, M. K. and Mai, A. T. (2018). Banking Characteristics of millennials. *Journal of Economic Cooperation and Development, Forthcoming.*

Casaló, L. V., Flavián, C. and Guinalíu, M. (2008). The role of satisfaction and website usability in developing customer loyalty and positive word-of-mouth in the e-banking services. *International Journal of Bank Marketing, 26*(6), 399–417.

Chatterjee, D., Grewal, R. and Sambamurthy, V. (2002). Shaping up for e-commerce: institutional enablers of the organizational assimilation of web technologies. *MIS Quarterly,* 65–89.

Chen, S. C. and Dhillon, G. S. (2003). Interpreting dimensions of consumer trust in e-commerce. *Information Technology and Management, 4*(2–3), 303–318.

Chu, P. Y., Lee, G. Y. and Chao, Y. (2012). Service quality, customer satisfaction, customer trust, and loyalty in an e-banking context. *Social Behavior and Personality: An International Journal, 40*(8), 1271–1283.

Corbitt, B. J., Thanasankit, T. and Yi, H. (2003). Trust and e-commerce: a study of consumer perceptions. *Electronic Commerce Research and Applications, 2*(3), 203–215.

Daniel, E., Wilson, H. and Myers, A. (2002). Adoption of e-commerce by SMEs in the UK: towards a stage model. *International Small Business Journal, 20*(3), 253–270.

Daniel, E. M. and Grimshaw, D. J. (2002). An exploratory comparison of electronic commerce adoption in large and small enterprises. *Journal of Information Technology, 17*(3), 133–147.

De Blasio, G. (2008). Urban – rural differences in internet usage, e-commerce, and e-banking: evidence from Italy. *Growth and Change, 39*(2), 341–367.

Dinev, T. and Hart, P. (2006). An extended privacy calculus model for e-commerce transactions. *Information systems research, 17*(1), 61–80.

Drigă, I. and Isac, C. (2014). E-banking services – features, challenges and benefits. *Annals of the University of Petroşani, Economics, 14*(1), 41–50.

Dyck, A., Morse, A. and Zingales, L., 2010. Who blows the whistle on corporate fraud?. *The Journal of Finance, 65*(6), 2213–2253.

Eastin, M. S. (2002). Diffusion of e-commerce: an analysis of the adoption of four e-commerce activities. *Telematics and informatics, 19*(3), 251–267.

Egger, F. N. (2001, June). Affective design of e-commerce user interfaces: How to maximise perceived trustworthiness. In *Proceeding International Conference Affective Human Factors Design* (pp. 317–324).

Federal Reserve Board. (2016, March). Consumers and Mobile Financial Services 2016. Retrieved September 19, 2018, from www.federalreserve.gov/econresdata/consumers-and-mobile-financial-services-report-201603.pdf.

Gefen, D. (2000). E-commerce: the role of familiarity and trust. *Omega, 28*(6), 725–737.

Gibbs, J., Kraemer, K. L. and Dedrick, J. (2003). Environment and policy factors shaping global e-commerce diffusion: A cross-country comparison. *The Information Society, 19*(1), 5–18.

Gibbs, J. L. and Kraemer, K. L. (2004). A cross-country investigation of the determinants of scope of e-commerce use: an institutional approach. *Electronic Markets, 14*(2), 124–137.

Grandón, E. E. and Pearson, J. M. (2004). Electronic commerce adoption: an empirical study of small and medium US businesses. *Information & Management, 42*(1), 197–216.

Gross, M. B., Hogarth, J. M. and Schmeiser, M. D. (2012). Use of financial services by the unbanked and underbanked and the potential for mobile financial services adoption. *Federal Reserve Bulletin, 98*(4), 1–20.

Gunasekaran, A., Marri, H. B., McGaughey, R. E. and Nebhwani, M. D. (2002). E-commerce and its impact on operations management. *International Journal of Production Economics*, *75*(1–2), 185–197.

Hertzum, M., Jørgensen, N. and Nørgaard, M. (2004). Usable security and e-banking: Ease of use vis-a-vis security. *Australasian Journal of Information Systems*, *11*(2).

Hong, W. and Zhu, K. (2006). Migrating to internet-based e-commerce: Factors affecting e-commerce adoption and migration at the firm level. *Information & Management*, *43*(2), 204–221.

Huang, H., Tan, J. and Liu, L. (2009, June). Countermeasure techniques for deceptive phishing attack. In *2009 International Conference on New Trends in Information and Service Science* (pp. 636–641). IEEE.

Jalal, A., Marzooq, J. and Nabi, H. A. (2011). Evaluating the impacts of online banking factors on motivating the process of e-banking. *Journal of Management & Sustainability*, *1*, 32.

Liao, Z. and Cheung, M. T. (2002). Internet-based e-banking and consumer attitudes: an empirical study. *Information & Management*, *39*(4), 283–295.

Loonam, M. and Loughlin, D. (2008). An observation analysis of e-service quality in online banking. *Journal of Financial Services Marketing*, *13*(2), 164–178.

Parmar, B. (2012). Protecting against spear-phishing. *Computer Fraud & Security*, 2012(1), 8–11.

Mattila, M., Karjaluoto, H. and Pento, T. (2003). Internet banking adoption among mature customers: early majority or laggards?. *Journal of Services Marketing*, *17*(5), 514–528.

Suh, B. and Han, I. (2002). The impact of customer trust and perception of security control on the acceptance of electronic commerce. *International Journal of Electronic Commerce*, *7*(3), 135–161.

Wigand, R. T. (1997). Electronic commerce: Definition, theory, and context. *The Information Society*, *13*(1), 1–16.

11 Feature selection-based data classification for stock price prediction using ant-miner algorithm

Saravanan Ramalingam and Pothula Sujatha

Introduction

Stock price prediction (SPP) is a hot research topic because it is useful for buyers and sellers of stocks. A massive amount of data is generated by the stock price market, and prices vary at each and every second. It is a kind of chaos system, in which a user can gain or lose their whole life savings in one second. Developing an accurate SPP model is very tedious since it is based on various factors like news, social media data, company production, government agreement, historical rate and a country's economics [1]. Investors and research communities spend time to devise methods to appropriately choose stocks with a favorable profit return to be part of an investment portfolio. Sellers are interested in whether the stock will increase or decrease after a particular time period. To select a company whose stock will rise, several investigation methods are designed, using present and previous data as well as basic information about the company. The status level of the company is identified from balance sheets and various ratios which will be helpful to forecast future stock process. Experts employ some statistical models using previous data to validate company's intrinsic value, like the Graham number, which is a commonly used prediction model [2]. But because of the nature of increased volatility in the present market scenario, it is not possible to identify companies which satisfy the Graham principles in a real-world scenario. Thus new and efficient SPP models are needed to adapt to the dynamic nature of the stock market [3]–[4]. Recent investment models with new ideas have been developed, and some of the classic methods became outdated. As financial literacy increases rapidly, the number of market players also increases. At the same time, earlier models are not applicable for variations in the stock market. The usage of algorithms in trading completely changed the stock market and made rapid changes easier. Machine learning (ML) allows developers to design models for SPP in a simple way using previous data.

In recent years, numerous approaches have been developed for SPP, such as genetic algorithms (GAs), neural networks (NN) and other artificial intelligence techniques [5]–[7]. These methods are used for data classification and are sometimes integrated with other methods for hybrid approaches [8]. Several researches have been carried out using evolutionary algorithms (EA) to do black-box investing. Feature selection plays a vital role in the data classification task. It is a process of selecting a subset of features and eliminating redundant and irrelevant features, thereby reducing the computational complexity of the system. The benefits of feature selection methods are shorter execution time, transparency, reducing the number of measurements and so on [26]–[29]. The intention of the feature selection algorithm is to lessen the computational complexity of the system and also to enhance the effectiveness of the learning algorithm. The feature selection can be modeled mathematically as a combinatorial optimization problem and the function selection as a dataset. The design variables are the addition (1) or the elimination (0) of the features. A comprehensive selection of features would assess numerous combinations (2^N, where N indicates the number of features). It is computationally complex; when the number of features is big, then it becomes impossible to calculate. Thus intelligent algorithms are needed where the feature selection will be carried out in a simpler way [30]–[31]. Some of the metaheuristic algorithms used for feature selection are ant colony optimization (ACO), GA, particle swarm optimization (PSO) and simulated annealing.

This paper analyzes the effect of feature selection approaches (GA and PSO) in the ant-miner based classification task of SPP. Ant miner is the famous technique which produces an effective solution for the classification problem. This paper proposes a novel SPP model of integrating feature selection methods in the ant-miner-based classification methods for effective SPP. The performance is validated by testing the proposed method against four datasets: the Dow Jones dataset and three datasets from Yahoo! Finance on a daily, weekly and monthly basis. The results are evaluated in terms of various measures like accuracy, sensitivity, specificity, kappa coefficient, F-score, false discovery rate (FDR) and false omission rate (FOR).

The rest of the chapter is arranged as follows. First, we summarize the existing SPP methods in detail. Then we explain the proposed model and discuss the dataset, performance metrics and the results.

Related work

The applications of ACO in diverse domains are discussed especially under classification problems. In addition, the existing SPP models are also explained here. In this section, previous works on SPP are discussed with the aims, methodologies, performance measures, benefits and some of the demerits. Several SPP techniques have been proposed based on mathematical frameworks, machine learning (ML) methods and bio-inspired algorithms. Here the existing works related to the proposed method are discussed in detail.

Uthayakumar et al. [9] used ACO-based rule induction to effectively predict whether or not financial firms will undergo bankruptcy. Alrasheedi et al. [10] designed a SPP model using different classifiers in the Saudi stock exchange during the period 2006–2013, and they utilized the Dow Jones dataset with a fivefold cross-validation technique. Nikola Milosevic et al. [11] developed an ML-based approach to investigate the equity's future price for a longer duration. This model precisely determines the rise of a company's value by 10% in a year. The intent in employing ML algorithms is to train the previous data; it is likely to forecast the stock prices and also to compute the ratio of movement for a specific time period. They compared different classifiers for SPP such as C4.5, support vector machine (SVM), logistic regression (LR), naïve Bayes (NB) and random forest (RF) in terms of precision, F-score and recall; RF attains better performance than other methods. Leung et al. [12] devised structural SVM (SSVM) to forecast the rise or fall of stock prices. It performs classification on complex inputs like the nodes of a graph. It enables SSVM to learn a prediction model for a complex graph input with multiple edges for every node. The resultant model is employed for SPP, in which the positive and negative class labels indicate the increase and decrease in stock prices. However, it uses threefold cross-validation to identify the actual value, and the SSVM parameter C is set to 1,000. It attains an accuracy of 78%, recommending the model has learned with no over-fitting. The obtained results show that the ML approach is a good choice to predict stock prices. Qiu et al. [13] used an artificial neural network (ANN) for SPP on the Japanese stock exchange. The intent of the study was to determine the next day's stock prices. To enhance the classification prediction, ANN was incorporated into the GA algorithm and presented as the GA-ANN method for effective SPP. GA is selected to improve the precision of ANN and also to avoid the convergence issue of the back propagation method. The attained results reported that the hybrid GA-ANN method achieves a hit rate of 81.27%, which is higher than the compared methods.

Zhiqiang Guo et al. [14] implemented a hybrid model which combines two-dimensional principal component analysis (2D-PCA) and radial basis function with neural network (RBF-NN) for SPP on the Shanghai stock market. They selected 36 stock market variables as input features, and a sliding window was used to get input data. Then 2D-PCA was used to lessen the dimensions of data and filter its intrinsic features. At last, RBF-NN used the data processed by 2D-PCA to find tomorrow's stock price. The simulation result proves that the proposed method outperforms multilayer perceptron (MLP). Khalid Alkhatib et al. [15] used a k-nearest neighbor (KNN) approach and non-linear regression method for SPP for six companies on the Jordanian stock exchange to assist sellers and buyers to make correct decisions. Based on the results, the KNN algorithm seems to be robust and achieves less error, and the prediction results are closer to the actual stock prices. Oguz Akbilgic et al. [16] introduce a hybrid RBF-NN (HRBF-NN) which combines regression trees, NN and RBF for SPP on the Istanbul stock exchange. HRBF-NN is highly useful to model complex non-linear relationships and dependencies between the stock indices. It produces better results even for non-linear data.

Zhiqiang Guo et al. [17] presented a SPP model using PCA, canonical correlation analysis and SVM. Initially, two features are filtered from the previous closing prices and 39 technical variables are attained from independent component analysis. Next, a canonical correlation analysis technique is used to integrate two types of features and filter intrinsic features to enhance prediction accuracy. Finally, SVM is used to forecast the next day's closing price. The experiment is carried out using the Dow Jones index, and the results indicate the proposed method is better than other methods. Though various methods have been devised for SPP, none of the methods utilizes ACO in the classification problem of SPP.

Feature selection-based SPP using ant-miner algorithm

This paper uses ant-miner algorithm for data classification purposes in SPP. To enhance the classification accuracy of the ant-miner algorithm, two feature selection methods, namely GA and PSO, are applied to select a subset of features and irrelevant features are eliminated from the applied dataset. The working operation of feature selection methods and the ant-miner algorithm is explained in the following subsections.

GA based feature selection

An important evolutionary algorithm (EA) for feature selection is GA, which is a stochastic way to optimize functions using the nature of genetics and biological evolution [18]. Generally, genes have the tendency to mutate over succeeding generations to adapt with the environment. GA operates on a population of individuals to develop better approximations. At every generation, a new population is developed by the process of selecting the individuals based on their level in the problem area and re-integrating them by the use of operators in natural genetics. The offspring may also perform a mutation, which results in a generation of individuals which are better adapted to the environment than the individuals that they were generated from, just as in natural adaptation. In SPP, every individual in the population indicates a predictive model. The number of genes is the number of available features in the dataset. Genes are binary values, which indicate the addition or deletion of certain features in the model. The number of individuals, or population size, selected for every application is set to ten by default ($N = 10$, where N is the number of features).

Initialization

The first step is to generate and initialize the individuals in the population. As GA is a stochastic optimization approach, the genes of the individuals are initialized randomly.

Fitness assignment

Once the population is initialized, the next step is to assign fitness to every individual. To verify the fitness function, the SPP model should be trained and then selection error is validated by the selection data. Generally, a high selection error indicates low fitness, and an individual with higher fitness has a higher probability of being chosen for recombination. The commonly used method to assign the fitness function is a rank-based method, where the selection errors of all the individuals are arranged and fitness is assigned to every individual based on the location in the individuals' rank and not on actual selection error.

Selection

When the fitness value is assigned to every individual, the selection operator selects the individuals that can be recombined for the upcoming generation. The individuals with the higher probability to survive are those more fitted to the environment. So, the selection operator chooses the individuals based on the fitness level. The number of chosen individuals is $N/2$. A stochastic sampling method named roulette wheel is used for the selection method. It places all the individuals on a roulette wheel, and the wheel is turned to randomly select individuals.

Crossover

The selection process selects only 50% of the population, and the crossover operator recombines the chosen individuals to generate a new population. This operator randomly selects two individuals and combines the features to produce four offspring for the new population, until the new population has a size similar to the previous one. The uniform crossover method makes a decision to select whether every offspring's features come from one parent or the other.

Mutation

The crossover operator can create offspring which are most likely the same as the parents. This might cause a next generation with low diversity. The mutation operator solves the issue by randomly altering the value of some features in the offspring. The entire process includes initialization, fitness assignment, selection and recombination. The mutation process is continued until a stopping condition is achieved.

PSO-based feature selection

PSO is an EA developed by Kennedy and Eberhart in 1995 [19]–[20]. PSO is inspired by social behaviors like birds flocking and fish schooling. The fundamental concept of PSO is the optimization of knowledge by social interaction in the population, where thinking is personal as well as social. PSO is based on the principle that every solution can be indicated as a particle in the swarm. Every particle has a position in the search space, which is defined by a vector $x_i = (x_{i1}, x_{i2}, \ldots, x_{iD})$, where D is the dimensionality of the search space. The particles move in the search space for searching the optimal solutions. Consequently, every particle has a velocity, which is represented as $v_i = (v_{i1}, v_{i2}, \ldots, v_{iD})$. During movement, every particle updates its position and velocity based on their experience and its neighbors. The best previous position of the particle is called personal best, or *pbest*, and the best position attained by the population is called global best, or *gbest*. Using *pbest* and *gbest*, PSO searches the optimal solutions by the updating of velocity and the position of every particle based on equations 11.1 and 11.2 [21].

$$x_{id}^{t+1} = x_{id}^{t} + v_{id}^{t+1} \tag{11.1}$$

$$v_{id}^{t+1} = w * v_{id}^{t} + c_1 * r_1 * \left(p_{id} - x_{id}^{t} \right) + c_2 * r_2 * \left(p_{gd} - x_{id}^{t} \right) \tag{11.2}$$

where t denotes the tth iteration in the evolutionary process, $d \in D$ indicates the dth dimension in the search space, w is inertia weight, c_1 and c_2 are acceleration constants, r_1 and r_2 are random values uniformly distributed in $[0, 1]$, and p_{id} and p_{gd} represent the elements of *pbest* and *gbest* in the dth dimension. The velocity is limited by a fixed maximum velocity, vmax, and $vt + 1id \epsilon [-vamx, vmax]$. The algorithm terminates when a predefined condition is satisfied, which can be a better fitness value or a fixed maximum number of iterations.

Ant-miner based SPP

When the features are selected using GA and PSO, the ant-miner algorithm is employed for SPP. In the ant-miner algorithm, the ants discover the shortest route from source to destination. The ants choose the feasible paths on the basis of a probability function. It is derived by the quantity of pheromone along the path and a heuristic function. When the ants visit all feasible paths, the path (because of positive feedback mechanism) with higher amounts of pheromone and the heuristic value with higher possibility will be chosen. When an ant chooses a path, the pheromone value will start to increase. When a sufficient number of ants chooses a single path, it will become a candidate rule and then a discovered rule once the quality is good enough.

Structural representation of classification rule induction

ACO is generally based on the foraging nature of real ant colonies. An attribute is defined as *Attribute*$_i$ where i denotes the series of the attribute. Va_{ij} represents the non-continuous attribute value, where i and j represents the series of the attribute. The next level of the attribute belongs to class, and the value of class is represented as CL_k, where k is the series value in the class. The ant initially starts from an artificial nest as a source and selects a value for every attribute. Once it visits all attributes, it chooses a value for the class and consumes the food as a destination. For rule discovery, enough ants should take the same path, which is described in the following section.

Rule structure

The classification based rule structure of ant-miner algorithm is given as: IF <antecedent > THEN <descendant>.

Rule generation

Sequential covering approach is utilized by ant-miner algorithm to identify the list of classification rules [22]. Initially, the number of discovered rules in the rule list is fixed as zero and the training set contains the discovered rules. When the classification rules are discovered for every repetition of WHILE loop parallel to a number of executions of the REPEAT-UNTIL loop, the rule will be shifted to classification rule list and eliminated from the training set [23]. This process repeats until the maximum threshold value crosses the number of uncovered training cases. At the start, Ant_t begins with zero rules and incremented by one term to its existing partial rule until one of the following conditions is satisfied:

1   Any value lesser than predefined threshold can be appended to the rule.
2   When the ants utilize all the attributes earlier, rule generation will be terminated. The artificial ants select an attribute value to create rules by the probability function, equation 11.3.

$$P_{xy} = \frac{\eta_{xy}.\tau_{xy}}{\sum_{x=1}^{a}(z_x).\sum_{y=1}^{b}\left(\eta_{xy}.\tau_{xy}(t)\right)} \tag{11.3}$$

where P_{xy} represents the likelihood function, η_{xy} represents the value of a problem-dependent heuristic function and $\tau_{xy}(t)$ represents the amount of pheromone at iteration t.

Rule pruning

When an ant generates a rule, the rule pruning process will be called. The pruning process removes the unwanted rules produced by the ants in each step. It enhances the quality of a rule created by the ants and the rules will be simple. Equation 11.4 gives the rule quality between zero and one: $0 \leq Q \leq 1$.

$$Q = \frac{TP}{(TP+FN)} * \frac{TN}{(FP+TN)} \tag{11.4}$$

where TP = true positive, TN = true negative, FP = false positive and FN = false negative.

Pheromone updating

Pheromone updating indicates the amount of ant pheromone evaporated in the real world. Artificial ants execute the pheromone updating procedure to identify the simpler rules [24]. Because of the positive feedback mechanism, the mistakes of the heuristic measure will be rectified and results in improved classification accuracy; this is shown in equation 11.5.

$$\tau_{xy}(t=0) = \frac{1}{\sum_{x=1}^{a}bx} \tag{11.5}$$

where *a* is the 'n' number of attributes, b_x is the possible values that associated attribute a_i.

Performance analysis

The performance of the feature selection–based ant-miner classification algorithm is tested and results are validated in this section. The dataset employed, metrics used and the obtained results are discussed in the following subsections. The results are compared between ant-miner algorithm without feature selection, ant-miner algorithm with GA-based feature selection and ant-miner algorithm with PSO-based feature selection.

Parameter setting

The parameters used in this experiment are tabulated in Table 11.1.

Dataset description

The performance is evaluated using a set of four datasets including a benchmark Dow Jones index dataset [25]. To ensure the effectiveness of the proposed methods, datasets are also gathered from Yahoo! Finance on a daily, weekly and monthly basis. These four datasets are used to assess the performance of the inclusion of feature selection methods in the classification task of SPP. The benchmark Dow Jones index dataset contains 750 instances and 16 attributes. The remaining three datasets hold 750 instances with seven attributes, namely date, opening price, high price, low price, closing price, adjusted close price and stock volume. The details of the applied four datasets are given in Table 11.2.

Performance measures

The performance measures used in this experiment for comparison purposes are accuracy, sensitivity, specificity, kappa coefficient, F-score, FOR and FDR. The measures are tabulated in Table 11.3 along with their mathematical formulas.

Result analysis

The obtained classification results of the three methods (ant-miner algorithm without feature selection, ant-miner algorithm with GA-based feature selection

Table 11.1 Parameter setting

Parameter	Variable	Value
Number of ants	Number_of_ants	2,000
Minimum number of cases	Min_cases_per_rule	5
Maximum number of uncovered cases	Max_uncovered_cases	10
Number of converged rules	No_of_rules_converge	10

Table 11.2 Dataset description

S. No	Dataset	No. of Instances	No. of Features	No. of Classes
1	Dow Jones Index	750	16	2
2	YahooFinance_2007_2017_Daily	2,755	7	2
3	YahooFinance_2007_2017_Weekly	133	7	2
4	YahooFinance_2007_2017_Monthly	572	7	2

Table 11.3 Performance measures

Parameter	Formula
Accuracy	$(TP + TN)/(TP + FP + FN)$
Sensitivity	$TP/(TP + FN)$
Specificity	$TP/(TN + FP)$
F-score	$2TP/(2TP + FP + TN)$
FOR	$\dfrac{FN}{FN + TN}$
FDR	$\dfrac{FP}{FP + TP}$
Kappa coefficient	$\dfrac{Obs.Agreement - Exp.Agreement}{100 - Exp.Agreement}$

and ant-miner algorithm with PSO-based feature selection) are compared against a same set of four datasets and are tabulated in Tables 11.4–11.6. The results obtained by the three methods for the four datasets are illustrated in Figures 11.1–11.4. From these figures, it is clear that PSO-based feature selection outperforms the other methods.

Table 11.4 shows the attained classification results of three methods under several performance metrics for the Dow Jones dataset. The table shows that the ant miner with no feature selection (i.e. none) selects all features (16 features) and then performs the classification task for SPP. But GA and PSO remove unwanted features and select only three and four features out of 16, respectively. The reduction in number of features allows the PSO and GA to perform classification at a faster rate with less computational complexity. From the table, it is clear that the PSO-based feature selection attains a maximum accuracy of 96.54, which is much higher than the classical ant-miner method. Though GA attains an accuracy of 90.4, it fails to achieve better results than PSO. The kappa value of 100 represents the measure of agreement between experts and models; a value closer to 100 indicates better system performance. The kappa values of PSO and GA are 92.16 and 80.28, respectively, whereas the ant-miner method attains a worse kappa value of 43.93. The classification accuracy of ant miner is 72, which is less than the other

Table 11.4 Classification results with Dow Jones index dataset

Method	Selected Features	Accuracy	Sensitivity	Specificity	F-score	FOR	FDR	Kappa
None	All	72	73.52	70.73	70.42	23.68	32.43	43.93
GA	7,10,13,17	90.4	87.42	92.59	88.53	9.09	10.32	80.28
PSO	17,16,10	96.54	92.87	98.76	95.54	5.43	1.63	92.16

Table 11.5 Classification results with YahooFinance_2007_2017_Daily dataset

Method	Selected Features	Accuracy	Sensitivity	Specificity	F-score	FOR	FDR	Kappa
None	All	87.29	81.30	91.69	84.41	13.01	12.22	73.72
GA	1,2,4,7,5	89.65	86.71	91.76	87.51	9.42	11.67	78.68
PSO	2,4,6	91.57	91.31	91.76	90.06	6.36	11.14	82.76

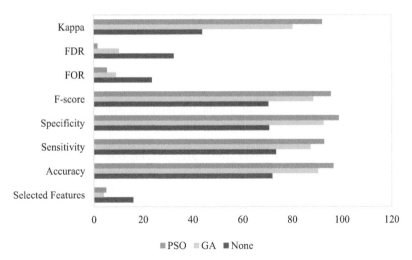

Dow Jones Index Dataset

■ PSO   ■ GA   ■ None

Figure 11.1 Comparative analysis of classification results for Dow Jones dataset

two methods, and proves that the classification accuracy of Dow Jones dataset can be significantly increased by the use of the feature selection method prior to classification.

Table 11.5 shows the attained classification results of three methods under several performance metrics for the Yahoo! Finance daily dataset in the years 2007–2017. Similar to the Dow Jones dataset, the table shows that the ant miner selects none of the features and then performs the classification task for SPP using the available seven features. But GA and PSO remove unwanted features and select

Table 11.6 Classification results with YahooFinance_2007_2017_Monthly dataset

Method	Selected Features	Accuracy	Sensitivity	Specificity	F-score	FOR	FDR	Kappa
None	All	87.41	88.23	86.20	89.28	16.66	9.63	74.03
GA	6,2,1,7,5	89.51	91.46	86.88	90.90	11.66	9.63	78.51
PSO	1,6,3	90.90	93.90	86.88	92.21	8.62	9.41	81.29

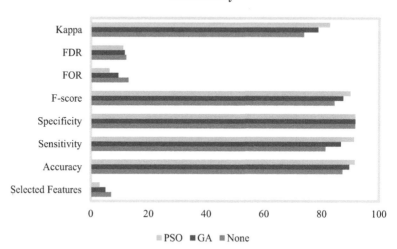

Figure 11.2 Comparative analysis of classification results for daily dataset

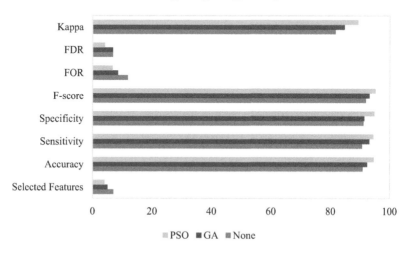

Figure 11.3 Comparative analysis of classification results for weekly dataset

Yahoo Finance_2007_2017_Monthly Dataset

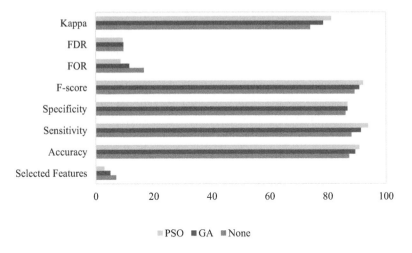

Figure 11.4 Comparative analysis of classification results for monthly dataset

only five and three features out of seven, respectively. The reduction in number of features enables PSO and GA to perform classification at a faster rate with less computational complexity. From the table, it is clear that the highest level of accuracy (91.57) is achieved by PSO, which is higher than the classical ant-miner and GA methods. Though GA attains an accuracy of 89.65, it fails to achieve better results than PSO. The kappa value of 100 represents the exact measure of agreement between experts and models; a value closer to 100 indicates better system performance. The kappa values of PSO and GA are 82.76 and 78.68, respectively, whereas the ant-miner method attains a worse kappa value of 73.72. The classification accuracy of ant miner is 87.29, which is less than the other two methods, which proves that the classification accuracy of the Yahoo! Finance daily dataset can be significantly increased by the use of the feature selection method prior to classification.

Table 11.6 shows the attained classification results of three methods under several performance metrics for the Yahoo! Finance weekly dataset in the years 2007–2017. The table shows that the ant miner selects none of the features and then performs the classification task for SPP using the available seven features. But, GA and PSO remove unwanted features and select only five and four features out of seven, respectively. The reduction in number of features enables PSO and GA to perform classification at a faster rate with less computational complexity. From the table, it is clear that the highest level of accuracy (94.73) is achieved by PSO, which is higher than the classical ant-miner and GA methods. Though GA attains an accuracy of 92.48, it fails to achieve better results than PSO. The kappa values of PSO and GA are 89.35 and 84.76, respectively, whereas the ant-miner method attains a worse kappa value of 81.65. The classification accuracy of ant miner is 90.97, which is less than the other two methods, which proves that the classification accuracy of the Yahoo! Finance daily dataset can be significantly increased by the use of the feature selection method prior to classification.

Table 11.6 shows the attained classification results of three methods under several performance metrics for the Yahoo! Finance monthly dataset in the year 2007–2017. Similar to the weekly dataset, the table shows that the ant miner selects none of the features and then performs the classification task for SPP using the available seven features. But GA and PSO remove unwanted features and select only five and three features out of seven, respectively. The reduction in number of features enables PSO and GA to perform classification at a faster rate with less computational complexity. From the table, it is clear that the highest level of accuracy (90.90) is achieved by PSO, which is higher than the classical ant-miner and GA methods. Though GA attains an accuracy of 89.51, it fails to achieve better results than PSO. The kappa value of 100 represents the exact measure of agreement between experts and models; a value closer to 100 indicates better system performance. The kappa values of PSO and GA are 81.29 and 78.51, respectively, whereas the ant-miner method attains a worse kappa value of 74.03. The classification accuracy of ant miner is 87.41, which is less than the other two methods, which proves that the classification accuracy of the Yahoo! Finance monthly dataset can be significantly increased by the use of feature selection method in prior to classification.

Conclusion

In this paper, the effect of feature selection method (GA and PSO) in SPP using ant-miner algorithm is investigated. The obtained results of ant-miner algorithm without feature selection, ant-miner algorithm with GA-based feature selection and ant-miner algorithm with PSO-based feature selection are compared in terms of various performance metrics such as accuracy, sensitivity, specificity, kappa coefficient, F-score, FOR and FDR. The empirical result proves that the PSO-based feature selection algorithm achieves better results than the other two methods. The results indicate that the inclusion of feature selection approaches significantly increases the classification accuracy of SPP. In future, this work can be extended by the application of bio-inspired algorithms to design an accurate SPP model.

References

[1] Nai-Fu Chen, S.A.R. and Roll, R., 1986. Economic forces and the stock market. *The Journal of Business*, 59(3): 383–403. Available: www.jstor.org/stable/2352710.
[2] Graham, B., 1949. *Intelligent investor*. s.l.: Harper.
[3] Barsky, R.B. and De Long, J.B., 1992. *Why does the stock market fluctuate?*, s.l.: National Bureau of Economic Research.

[4] Hendershott, T. and Moulton, P. C., 2011. Automation, speed, and stock market quality: The NYSE's hybrid. *Journal of Financial Markets*, 14(4): 568–604.

[5] Mahfoud, S. and Mani, G., 1996. Financial forecasting using genetic algorithms. *Applied Artificial Intelligence*, 10: 543–565.

[6] Tsang, E. and Markose, S., Er, H. 2005. Chance discovery in stock index option and future arbitrage. *New Mathematics and Natural Computation*, 1(3): 435–477.

[7] Wagman, L., 2003. Stock portfolio evaluation: An application of genetic-programming-based technical analysis. Genetic Algorithms and Genetic Programming at Stanford, 2003, pp. 213–220.

[8] Wang, J. J., Wang, J. Z., Zhang, Z. G. and Guo, S. P. 2012. Stock Index forecasting based on a hybrid model. *Omega*, 40(6): 758–766.

[9] Uthayakumar, J., Vengattaraman, T. and Dhavachelvan, P. 2017. Swarm intelligence based classification rule induction (CRI) framework for qualitative and quantitative approach: An application of bankruptcy prediction and credit risk analysis. *Journal of King Saud University-Computer and Information Sciences* (in press).

[10] Alrasheedi, M. and Alghamdi, A. 2014. Comparison of classification methods for predicting the movement direction of Saudi Stock Exchange Index, *Journal of Applied Sciences*, 14(16): 1883–1888.

[11] Milosevic, Nikola. Equity forecast: Predicting long term stock price movement using ML. *arXiv preprint arXiv*:1603.00751 (2016).

[12] Leung, Carson Kai-Sang, MacKinnon, R. K. and Wang, Y. 2014. A ML approach for stock price prediction. In *Proceedings of the 18th International Database Engineering & Applications Symposium* (pp. 274–277). ACM.

[13] Qiu, M. and Song, Y. 2016. Predicting the direction of stock market index movement using an optimized artificial neural network model. *PloS One,* 11(5): e0155133.

[14] Guo, Zhiqiang, Wang, H., Yang, J. and Miller, David J. 2015. A stock market forecasting model combining two-directional two-dimensional principal component analysis and radial basis function neural network. *PloS One*, 10(4): e0122385.

[15] Alkhatib, Khalid, Najadat, H., Hmeidi, I., and Ali Shatnawi, M. K. 2013. Stock price prediction using k-nearest neighbor (knn) algorithm. *International Journal of Business, Humanities and Technology*, 3(3): 32–44.

[16] Akbilgic, Oguz, Bozdogan, H., and Erdal Balaban, M. 2014. A novel Hybrid RBF Neural Networks model as a forecaster. *Statistics and Computing*, 24(3): 365–375.

[17] Guo, Zhiqiang, Wang, H., Liu, Q., and Yang, J. 2014. A feature fusion based forecasting model for financial time series. *PloS One*, 9(6): e101113.

[18] Saidi, Rania, Waad Bouaguel Ncir, and Essoussi, N. 2018. Feature selection using genetic algorithm for big data. In *International Conference on Advanced Machine Learning Technologies and Applications* (pp. 352–361). Springer, Cham, 2018.

[19] Kennedy, J. and Eberhart, R. 1995. Particle swarm optimization. In *Proceeding IEEE International Conference Neural Network*, 4: 1942–1948.

[20] Shi, Y. and Eberhart, R. 1998. A modified particle swarm optimizer. In *Proc.IEEE Int. CEC*: 69–73.

[21] Yong Zhang, Gong, Dun-wei, Sun, Xiao-yan and Guo, Yi-nan. 2017. A PSO-based multi-objective multi-label feature selection method in classification. *Scientific Reports*, 7(1) (2017): 376.

[22] Predawan, S., Kimpan, C. and Wutiwiwatchai, C., 2009. Thai Tone Recognition Using Ant Colony Algorithm. *2009 International Conference Information Management and Engineering*, 181–185. http://doi.org/10.1109/ICIME.2009.49.

[23] Nair, B.B., Mohandas, V.P., Sakthivel, N.R., 2011. Predicting stock market trends using hybrid ant-colony-based data mining algorithms: An empirical validation on the Bombay Stock Exchange. *International Journal of Business Intelligence and Data Mining*, 6: 362–381. http://doi.org/10.1504/IJBIDM.2011.044976.

[24] Wan, Y., Wang, M., Ye, Z., Lai, X., 2016. A feature selection method based on modified binary coded ant colony optimization algorithm. *Applied Soft Computing*, 49: 248–258. http://doi.org/10.1016/j.asoc.2016.08.011.

[25] Dow Jones dataset available at https://archive.ics.uci.edu/ml/datasets/dow+jones+index.

[26] Shankar, K., Elhoseny, M., Lakshmanaprabu, S. K., Ilayaraja, M., Vidhyavathi, R. M., Elsoud, Mohamed A. and Alkhambashi, Majid. Optimal feature level fusion based ANFIS classifier for brain MRI image classification. *Concurrency and Computation: Practice and Experience,* August 2018. (DOI: https://doi.org/10.1002/cpe.4887)

[27] Karthikeyan, K., Sunder, R., Shankar, K., Lakshmanaprabu, S. K., Vijayakumar, V., Elhoseny, M. and Manogaran, G. Energy consumption analysis of Virtual Machine migration in cloud using hybrid swarm optimization (ABC – BA), *The Journal of Supercomputing*, First Online: 05 September 2018, (DOI: https://doi.org/10.1007/s11227-018-2583-3)

[28] Hassanien, Aboul Ella, Rizk-Allah, Rizk M., and Elhoseny, Mohamed. A hybrid crow search algorithm based on rough searching scheme for solving engineering optimization problems, *Journal of Ambient Intelligence and Humanized Computing*, June 2018, https://doi.org/10.1007/s12652-018-0924-y.

[29] Rahmani Hosseinabadi, Ali Asghar, Vahidi, Javad, Saemi, Behzad, Kumar Sangaiah, Arun, Elhoseny, Mohamed. 2018. Extended genetic algorithm for solving open-shop scheduling problem, *Soft Computing*, Springer, April 2018 (https://doi.org/10.1007/s00500-018-3177-y)

[30] Metawa, N., Kabir Hassana, M. and Elhoseny, M. 2017. Genetic algorithm based model for optimizing bank lending decisions. *Expert Systems with Applications*, 80: 75–82 (https://doi.org/10.1016/j.eswa.2017.03.021)

[31] Elhoseny, Mohamed, Tharwat, Alaa, Farouk, Ahmed and Hassanien, Aboul Ella. 2017. K-coverage model based on genetic algorithm to extend WSN lifetime. *IEEE Sensors Letters*, 1(4): 1–4.

12 The value of simulations characterizing classes of symbiosis

ABCs of formulation design

K. Basaid, B. Chebli, J. N. Furze, E. H. Mayad and R. Bouharroud

Introduction

Crop protection is constantly faced with challenges to satisfy customer demands while promoting crop productivity and ecosystem maintenance. The search for new control agents is continuous, as crops are subject to attacks by many pathogens (fungi, bacteria, nematodes, insects, etc.). This work investigates plant essential oils for biological control of post-harvest fungi. There are two approaches to deal with this subject. The first one is a basic approach, in which we characterize plants with laboratory tests and use statistics for results interpretation. It is a limited approach in terms of mastery of interaction on different levels. Hence we complement it with a second one, a functional approach, where we apply mathematical theory and modeling to plant systems. This approach enables understanding how the relationship works between different classes subject to the investigation, and within the classes, thus applying this knowledge for simulation of the synergy between different members of the ecosystem community. First we discuss the use of plant essential oils for crop protection, and then we give an overview of fuzzy modeling and the importance of its application in this context. Finally, we discuss the choice of plant species used and we state objectives of the investigation.

The frequent applications of commercial fungicides can cause negative effects on the environment and food safety (Norman, 1988), particularly in the post-harvest period because of the short time between treatment and consumption (Chebli et al., 2004). Furthermore, it leads to the development of resistance in pathogens. An example is the fungus *Botrytis cinerea*, which became resistant to specific fungicides in a short period of time (Elad et al., 1992). This species belongs to *Botrytis*, a highly diverse genus, with more than 30 species that are necrotrophic pathogens decaying infected plant tissues (Fillinger and Elad, 2016). The most common and the most important species of this genus is *B. cinerea* Pers., a ubiquitous pathogen responsible for losses in over 200 crop species worldwide in pre- and post-harvest (Williamson et al., 2007; Chebli et al., 2003).

The negative perception of synthetic fungicides used for control of *Botrytis cinerea* drives attention towards natural alternatives, mainly among plant essential oils (Sharma and Tripathi, 2008). These phyto-compounds are biodegradable, cause minimal effects on nontarget organisms and delay the occurrence of resistance in pests (Isman, 2000). Thus inhibition of *B. cinerea* by essential oils is the subject of several studies (Cid-Pérez et al., 2016; Fraternale et al., 2016; Ouadi et al., 2017; Xueuan et al., 2018; Banani et al., 2018), which provides motivation to investigate other plant species. Further, we wanted to expand the knowledge on these novel species by characterizing plant systems with mathematic precision, ultimately enabling fuzzy logic methods.

For Sugeno and Yasukawa, fuzzy modeling enabled formation of a system model with the help of a description language, which consists of fuzzy logic with fuzzy predicates, where in a larger sense fuzzy modeling was used to describe system behavior in a qualitative way, using a natural language. In a constricted sense, they referred the use of fuzzy modeling to describing systems with fuzzy quantities (Sugeno and Yasukawa, 1993). Since then, fuzzy logic has evolved. It is underpinned by set theory and, as such, is used to mathematically describe the quantification of complex design problems and form control strategies on which one bases a rule structure (Furze, 2014).

Seminal thinking about fuzzy modeling started with the initial papers of Zadeh, as he described a human action qualitatively using fuzzy algorithms (Zadeh, 1968). Later in 1973, he published his most remarkable work related to qualitative modeling, which established the foundation for fuzzy control. In this paper, he introduced the concept of linguistic variables and proposed to use fuzzy IF-THEN rules to formulate human knowledge (Zadeh, 1973). Motivated by these ideas of "fuzzy algorithm" and "linguistic analysis", Mamdani applied fuzzy logic initially to control (Mamdani, 1974). Mamdani and Assilian established the basic framework of the fuzzy controller (Mamdani and Assilian, 1975). By the 1980s, Takagi, Sugeno and Kang proposed another fuzzy system whose inputs and outputs are real-valued variables (Takagi and Sugeno, 1985; Sugeno and Kang, 1988).

Identification procedures and advanced knowledge investigation require the development of highly precise models (Lughofer, 2011). Mathematical process models give a better insight and furthermore understanding of the complete system, some specific system behaviors or process properties through analyzing the model structures and parameters. Such knowledge gains from process models may be taken as cornerstones for future design decisions (Lin and Segel, 1974). In the context of biological modeling, areas of study that need to be detailed are plant systems and process operations.

The plant species *Senecio glaucus* ssp. *coronopifolius* and *Ridolfia segetum* (L.) Moris are species of south-west Moroccan flora. They combine competitive and rudural strategies with traits of stress tolerance. The chemical profile of these species has been described (Cabral et al., 2015; Pooter et al., 1986). There are no studies on biological activities of *S. glaucus* ssp. *coronopifolius*, and only few studies on biological properties of *R. segetum*, which reported the antioxidant, antibacterial, anti-inflammatory and HIV-1-inhibiting activities of the essential oil (Cabral et al., 2015; Jabrane et al., 2010; Bicchi et al., 2009). No reports were made on the antifungal activity of these oils or on simulation of their interaction with fungi. Application of mathematical theory and modeling enables simulation of the synergy between plants and different life forms. Furthermore, it will help in the design of a function that works in different scenarios, allowing prediction of inhibition with high accuracy.

Our objectives are to show the antifungal potential of essential oils of *S. glaucus* ssp. *coronopifolius* and *R. segetum* against *Botrytis cinerea* with laboratory experimentation. Further, we apply mathematical approaches towards simulation of synergy in different scenarios.

Materials and methods

In this section, the laboratory tests are detailed. These include extraction of essential oils from plants, isolation of the fungus and application of oils to it by means of two methods. Further, we introduce fuzzy inference systems, mainly the Mamdani system, which we used in this work.

Experimental work

Collection of plants and extraction of essential oils

The collection of *Senecio glaucus* ssp. *coronopifolius* and *Ridolfia segetum* plants from the region of Souss Massa Darâa (south-west of Morocco) was done at the flowering stage. After drying, whole plants of *S. glaucus* ssp. *coronopifolius* (stems, leaves, flowers and roots) and aerial parts (stems, leaves, flowers and fruits) of *R. segetum* were used for extraction of essential oils. The method used was hydro-distillation via a Clevenger apparatus for four hours. The average yield of essential oils was 0.08% for *S. glaucus* ssp. *coronopifolius* and 0.05% for *R. segetum*.

Fungal isolation

Isolation of *Botrytis cinerea* was performed from rotten beans infected by the fungus, by transferring small fragments of the rotten beans to sterile petri dishes, filled with fresh potato dextrose medium (PDA) mixed with antibiotic, to prevent bacteria growth. Plates were put in incubation at 25°C for seven days. After a series of purifications, we attained a pure culture of the fungus, which was maintained on PDA at 4°C.

Antifungal activity assay

Two methods were used to test antifungal activity of essential oils against the fungus: the poisoned food method (PF) (Rhayour et al., 2003), and volatile phase method (VF) (Soylu et al., 2010), with some variations. The PF technique requires the mixture of essential oils, emulsified by Tween 80 with PDA, right before being emptied into petri dishes (9 cm diameter). The concentrations tested ranged from 0.25 to 16 μl/ml. The controls were filled with PDA alone. Moreover, the tested fungi was inoculated using a 6 mm mycelial plug from a seven-day-old culture. Plates were incubated for seven days at 25±2°C (Figure 12.1a).

In VF, Petri dishes (9 cm) filled with PDA and inoculated with the fungus were turned upside down. The inverted lids had sterile Whatman No. 1 filter paper discs, which contained different amounts of essential oils (10, 20, 40, 80, 160, 320, and 640 µl/disc). The controls had sterilized paper filter discs impregnated with distilled water. Plates were incubated at 25±2°C for seven days (Figure 12.1b).

According to the two techniques described above, for each treatment three replicate plates were inoculated. Fungus growth was determined by measuring the surface covered by the mycelium on petri dishes. Antifungal activity of essential oils is expressed by inhibition percentage of mycelial growth (I%) and evaluated according to equation 12.1 (Pandey et al., 1982):

$$I\% = (Dt\text{-}Di)*100 \qquad (12.1)$$

where Dt represents mycelial growth diameter in control petri plates and Di represents mycelial growth diameter in treated petri plates.

Statistical analysis

Growth inhibition data were submitted to analysis of variance (ANOVA). Mean values were compared using Tukey's test (also called Tukey's Honest Significant Difference, or HSD) ($p < 0.05$). The latter compares specific groups' means with each other to generate different ones. For each pair of means, we determine the HSD value using equation 12.2:

$$HSD = (Mi\text{-}Mj)/\sqrt{(MSw/Nh}) \qquad (12.2)$$

where Mi and Mj are two different means, and their subtraction is the difference between the pair of means; MSw represents the mean square within the pair of means; and N stands for number in the group.

Step 1 of Tukey's test is to perform the ANOVA test. If we find a significant F, we move to step 2, where we choose two means from the ANOVA output. Step 3

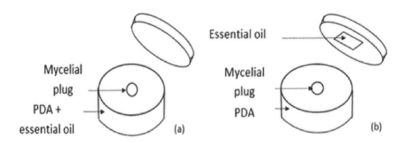

Figure 12.1 Poisoned food and volatile phase method

is to calculate the HSD value for the Tukey test with the help of equation 12.2, shown above.

Step 4 is to find the HSD value among values of Tukey's critical value table. Step 5 is to compare the HSD value calculated in step 3 with the critical value found in step 4. The two means are considered significantly different if the calculated value is bigger than the critical value.

Further, we calculated the IC_{50} value (concentration required for inhibition of 50% of mycelial growth). We calculated log values of essential oil concentrations, then determined their probit with inhibition percentage of the test fungus, using the linear regression. The calculated IC_{50} value represented the average of three replications.

Fuzzy inference systems

Fuzzy systems have a knowledge base consisting of fuzzy IF-THEN rules. These are statements that are characterized by continuous membership functions (Wang, 1997). In terms of inference process, there are two main types of fuzzy inference systems (FIS): the Mamdani type (Mamdani, 1974) and the TSK type (Takagi and Sugeno, 1985; Sugeno and Kang, 1988). In this investigation, we will focus on application of the first type.

Bayesian inference systems show the use of fuzzy logic to qualify modeling frameworks. The Mamdani FIS is widely used, it gives practical results with a simple structure and it has an intuitive and interpretable nature in its rule base (Mendel, 2001). They can be defined with fuzzy inference and fuzzy consequence. A basic configuration of a Mamdani system is shown in Figure 12.2.

A collection of IF-THEN rules (fuzzy rules) constitutes the rule base. The inputs and outputs are fuzzy sets. The fuzzy inference engine uses fuzzy logic principles in combining fuzzy rules, to plot from fuzzy input sets (U C Rn) to fuzzy output sets (V C R) (Wang, 1997). The same system has been used to calculate plant strategies (Furze et al., 2017; Grime, 2006; Furze et al., 2013b). If the dashed feedback line in Figure 12.2 exists, the system becomes the so-called

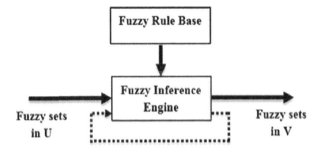

Figure 12.2 Process of Mamdani system (Wang, 1997)

Note: Input space U C Rn, Output space V C R.

fuzzy dynamic system (Wang, 1997). Application of a Mamdani system in this investigation is shown in Figure 12.3.

This system has low resolution input. A1(1) and A1(2) are essential oils of *Senecio glaucus* and *Ridolfia segetum*, respectively. A2 is the fungus *Botrytis cinerea*. The interaction between the input variables is inhibition. The latter varies from no inhibition B(1), to medium B(2) or maximum inhibition B(3) depending on oil doses. This system can be projected to various scenarios. Accordingly, candidates of input variables are different types of oils and different plants, and may also include different temperatures or environments. The outputs are a fungus, different types of fungus or different types of life forms (insects, bacteria, nematodes, etc.). The operation describes the interaction.

Results

Antifungal activity of essential oils

Table 12.1 presents the effects of *Senecio glaucus* ssp. *coronopifolius* and *Ridolfia segetum* essential oils concentrations on mycelial growth after a seven-day incubation period of *Botrytis cinerea* at $25 \pm 2°C$, using the PF technique.

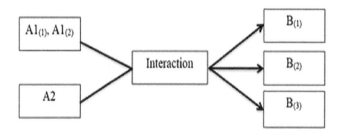

Figure 12.3 Qualitative determination of inhibition

Table 12.1 Inhibition percentages of *Botrytis cinerea* growth at various concentrations of the two essential oils using the PF technique

	Inhibition (%)	
Concentration (µl/ml)	*Senecio glaucus*	*Ridolfia segetum*
0	0.0 a	0.0 a
0.25	0.0 a	0.0 a
0.5	0.0 a	0.0 a
1	0.0 a	0.0 a
2	0.0 a	33 b
4	13 b	78 c
8	60 c	79 c
16	83 d	98 d

Values of the same column followed by the same letter are not different in a significant way according to Tukey's test ($p \leq 0.01$).

Inhibition percentage increases with concentration. Its maximum value reached 83% at 16 µl/ml for *Senecio glaucus* ssp. *coronopifolius*, and 98% at 16 µl/ml for *Ridolfia segetum*. IC_{50} values are 7.87 and 2.70 µl/ml, respectively (Figures 12.4 and 12.5).

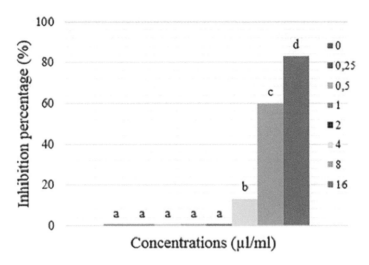

Figure 12.4 Inhibition percentages of *Senecio glaucus* ssp. *coronopifolius* essential oil against *Botrytis cinerea* at different concentrations using poisoned food method

Note: Values assigned the same letter do not differ significantly, according to Tukey's test ($p \leq 0.01$).

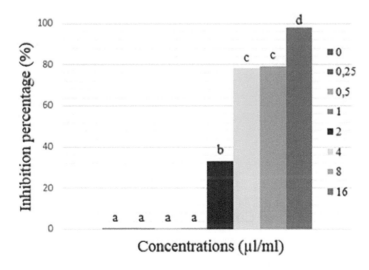

Figure 12.5 Inhibition percentages of *Ridolfia segetum* essential oil against *Botrytis cinerea* at different concentrations using poisoned food method

Note: Values assigned the same letter do not differ significantly, according to Tukey's test ($p \leq 0.01$).

Using the micro-atmosphere technique, the results of the effect of essential oil vapors on the growth of *Botrytis cinerea* mycelium after incubation at $25 \pm 2°C$ for seven days are presented in Table 12.2.

Values of the same column followed by the same letter are not significantly different according to Tukey ($p \leq 0.01$).

Maximum inhibition achieved by *Senecio glaucus* ssp. *coronopifolius* is 68% at 640 µl/disc, whereas *Ridolfia segetum* provided 80% inhibition of the fungus at the concentrations 640 µl/disc and 320 µl/disc (Figures 12.6 and 12.7).

Table 12.2 Inhibition percentages of *Botrytis cinerea* growth at various concentrations of the two essential oils using the VF technique

Concentration (µl/disc)	Inhibition (%)	
	Senecio glaucus	*Ridolfia segetum*
0	0.0 a	0.0 a
10	0.0 a	0.0 a
20	0.0 a	0.0 a
40	0.0 a	0.0 a
80	0.0 a	0.0 a
160	0.0 a	0.0 a
320	0.0 a	80 b
640	68 b	80 b

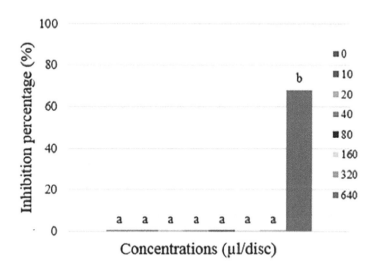

Figure 12.6 Inhibition percentages of *Senecio glaucus* ssp. *coronopifolius* essential oil against *Botrytis cinerea* at different concentrations using volatile phase method

Note: Values assigned the same letter do not differ significantly, according to Tukey's test (p ≤ 0.01).

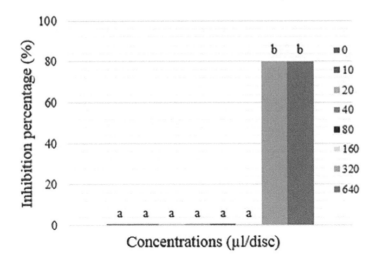

Figure 12.7 Inhibition percentages of *Ridolfia segetum* essential oil against *Botrytis cinerea* at different concentrations using volatile phase method

Note: Values assigned the same letter do not differ significantly, according to Tukey's test ($p \leq 0.01$).

The distribution of results obtained with the volatile phase method (VF) is a Poisson distribution, whereas the results obtained with the poisoned food method (PF) have a sigmoid distribution. This difference in distribution reflects on difference in resolution. Thus the PF method has low resolution and VF offers higher resolution. Hence for the following section, we will use results of VF method for application of a high resolution system.

Application of a Mamdani system

Figure 12.8 presents a higher resolution system based on the low resolution system showed in Figure 12.3. Given that chemical characterization of the oils used in this study has yet to be determined, we referred to references on similar oils to determine major compounds of oils, and to attribute to each major compound a percentage in which it is present in oils (Cabral et al., 2015; Pooter et al., 1986; Jabrane et al., 2010; Fleisher and Fleisher, 1996). The letters from A to G represent major compounds of essential oils (myrcene, dehydrofukinone, α-phyllandrene, p-cymene, terpinoline, dillapiol and myristin), and H stands for *Botrytis cinerea*. The quantity in oil attributed to each compound is the result of equation 12.3:

$$Q = P * OQ \tag{12.3}$$

where Q is the quantity in oil required for each compound, P is the percentage in which the compound is present in the oil and OQ is the quantity in oil applied to achieve inhibition.

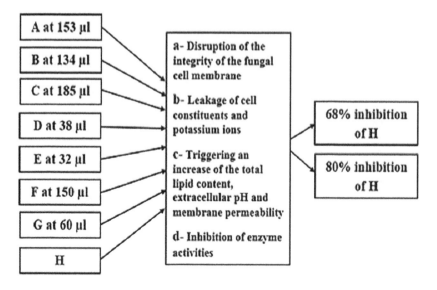

Figure 12.8 Identification of groups participating in inhibition

The percentages of major compounds in oils according to literature are as follows:

P (myrcene) = 24% (maximum value) (Pooter et al., 1986)
P (dehydrofukinone) = 21% (maximum value) (Pooter et al., 1986)
P (α-phyllandrene) = 58% (average of the range 53.0–63.3%) (Cabral et al., 2015)
P (p-cymene) = 12% (average of the range 8.4–15.2%) (Fleisher and Fleisher, 1996)
P (terpinoline) = 10% (average of the range 11.9–8.6%) (Cabral et al., 2015)
P (dillapiol) = 47% (maximum value) (Jabrane et al., 2010)
P (myristin) = 19% (maximum value) (Jabrane et al., 2010).

For values of quantity in oil (OQ), we will use those found with the VF method (high resolution): 640 μl for compounds myrcene and dehydrofukinone present in *Senecio glaucus* oil, and 320 μl for the rest of the compounds (α-phyllandrene, p-cymene, terpinoline, dillapiol and myristin), which are present in *Ridolfia segetum* oil.

The input of this system is major compounds of oils (A to G). The operation is different mechanisms of action of these compounds on the fungus (H). The outcome is the inhibition percentage of the fungus.

For a more simplified system structure, we propose the process presented in Figure 12.9.

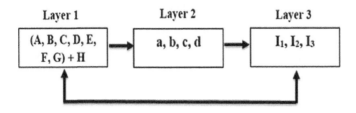

Layer 1	Layer 2	Layer 3
(A, B, C, D, E, F, G) + H	a, b, c, d	I_1, I_2, I_3

Figure 12.9 Process of inhibition

The input of the system are major compounds of oils (A to G). The operation is their mechanisms of action (a to d). The output of the system is inhibition of the fungus: I_1 is inhibition 1 (68% inhibition of H), I_2 is inhibition 2 (80% inhibition of H) and I_3 is no inhibition (0% inhibition of H). The arrow, which links layer 1 with layer 3, is indicative of a dynamic relation. The interaction between members of layer 1 creates feedback in the form of inhibition. Increase in layer 1 produces increase in layer 3, and a decrease in layer 3 has a feed forward to layer 1.

This process hints at equations 12.4, 12.5, and 12.6, by which we can design a formulation that is effective on the fungus.

$$A*153 + B*134 + H = 68\% \downarrow H \tag{12.4}$$

$$C*185 + D*38 + E*32 + F*150 + G*60 + H = 80\% \downarrow H \tag{12.5}$$

$$(A + B + C)*0 + (D + E + F + G)*0 + H = H \tag{12.6}$$

Equation 12.4 is built on the results obtained by *Senecio glaucus* essential oil. In the equation, the oil is replaced by its major compounds myrcene (A) and dehydrofukinone (B), multiplied by the quantities in which they can be used. With the addition of the fungus, the result for this equation is inhibition of the fungus by 68%.

Equation 12.5 is based on the results obtained by *Ridolfia segetum* essential oil. In the equation the oil is replaced by its major compounds α-phyllandrene (C), p-cymene (D), terpinoline (E), dillapiol (F) and myristin (G), multiplied by the quantities in which they can be used. With the addition of the fungus, the result for this equation is inhibition of the fungus by 80%.

Equation 12.6 combines major compounds of *Senecio glaucus* and *Ridolfia segetum* essential oils, all multiplied by 0. With the addition of the fungus, the result for this equation is the presence of the fungus, which indicates that in the absence of compounds of oils, there is no effect on the fungus.

Discussion

Oils of both species studied inhibit *Botrytis cinerea* and are optional antifungal agents. Yet *Ridolfia segetum* (L.) Moris provides greater inhibition of the fungus than *Senecio glaucus* ssp. *coronopifolius*. The methods used infer a difference in the chemical profiles of the oils. Major components of *S. glaucus* ssp. *coronopifolius* oil are myrcene (24%) and dehydrofukinone (21%) which gives the oil its distinct odor (Pooter et al., 1986). These two compounds have not been reported as antifungal against *B. cinerea*. Yet myrcene showed antifungal activity against *Fusarium oxysporum* (Sekine et al., 2007). A study by Silva et al. (2016) showed that dehydrofukinone had great antifungal activity against *Pycnoporus sanguineus* and *Gloeophyllum trabeum*.

Ridolfia segetum produces two types of oils: those that contain monoterpene hydrocarbons as major compounds (α-phellandrene, p-cymene, and terpinolene) (Cabral et al., 2015; Bicchi et al., 2009; Fleisher and Fleisher, 1996; Marongiu et al., 2007; Palá-Paúl et al., 2002; Palá-Paúl et al., 2005) and those that additionally contain phenylpropanoids like myristicin and dillapiol either as major compounds or in considerable amounts (Jabrane et al., 2010; Jannet and Mighri, 2007). P-cymene has previously showed good antifungal activity against *Botrytis cinerea* (Nabigol and Morshedi, 2011), and other fungi such as *Pythium ultimum, Fusarium acuminatum, Aspergillus niger, F. solani, Penicillium digitatum, R. solani, F. oxysporum, Verticillium dahlia, Alternaria mali, Aspergillus flavus, Aspergillus fumigatus* and *Penicillium sp.* (Sekine et al., 2007; Kordali et al., 2008; Marei and Abdelgaleil, 2018; Hammer et al., 2003). Terpinolene showed antifungal action against *Botrytis cinerea* (Yu et al., 2015) and other fungi such as *Aspergillus niger, Aspergillus flavus, Aspergillus fumigatus* and *Penicillium sp.* (Hammer et al., 2003). α-phellandrene is reported to have great antifungal activity against *Penicillium cyclopium* (Zhang et al., 2017). Myristicin was shown to have antifungal action on *Aspergillus flavus* and *Aspergillus ochraceus* (Valente et al., 2014), and dillapiol showed great antifungal activity against *Colletotrichum acutatum, Botryodiplodia theobromae* and *Clinipellis perniciosa* (Vizcaíno-Páez et al., 2016; DeAlmeida et al., 2009).

Given the diversity of molecules present in both essential oils, antifungal activity seems to result from a combination of several modes of action, involving different cellular targets. For example, α-phellandrene inhibits the growth of *Penicillium cyclopium* by acting on membrane permeability. That leads to leakage of cell constituents and increase of the total lipid content (Zhang et al., 2017). Marei and Abdelgaleil (2018) suggested that p-cymene gains antifungal activity by inhibiting enzyme activities, as it was shown to have inhibitory activity on pectin methyl esterase isolated from *Alternaria solani* and *Phytophthora infestans*, and cellulase isolated from *Phytophthora infestans* and *Aspergillus niger*. Dillapiol disrupts the integrity of cell membranes. Due to its lipophilic characteristic, it interferes with fatty-acid-chain constituents of the membrane (DeAlmeida et al., 2009).

Application of a basic Mamdani system is the initial step towards prediction of inhibition. To characterize the interaction going on between oils and fungi, we used a high-resolution Mamdani system. The system presented in Figure 12.8 shows possible mechanisms of action of major compounds of oils tested in fungus inhibition. This system has applications in crop protection. Figure 12.9 presents a simplified structure based on Figure 12.8, which gives a hint on the estimation of a recipe for a formulation (equations 12.3, 12.4 and 12.5), which can be applied against the fungus. This formulation would need to be further tested directly on plants infected with the fungus to ensure its effectiveness and lack of phytotoxicity on plants. Furthermore, plants in their ecosystem are not in seclusion; they are surrounded by other pathogenic life forms (insects, bacteria, nematodes, etc.) and are subject to their attacks, which opens up our minds to other scenarios where essential oils can have a key role in crop protection, thanks to their chemical diversity. Thus it is very interesting to test inhibitory effects of essential oil compounds on them to verify those ideas. I am currently exploring the effects of these oils on plant parasitic insects and nematodes, which enables the formation of broader functional scenarios that have ecosystem application.

Determination of rules for chemical profiles of oils and their outcome of inhibition on different life forms allows creation of formulas that possess several targets, which will promote plant growth and also benefit plant ecology, as essential oils don't have effects on nontarget organisms, which include beneficial organisms that live in symbiosis with plants in the ecosystem.

The next step after qualifying is quantifying (measurement) with quantifiable modeling structure, otherwise known as quantified hybrid systems that use higher mathematics. Higher mathematic approaches rely on centralized means and unit variance. Further, they make use of probability and distribution functions. Quantifiable modeling with defined variables and measured resolution produces accurate results, and they can be further expanded through other mathematic approaches with 100% certainty (Furze et al., 2013a). Application of higher mathematic methods involves the formation of a framework for synergy, in which case we are considering different memberships (plants, fungi, nematodes, insects, bacteria and the environment). Figure 12.10 shows a low resolution system that makes a framework for synergy in different scenarios between different life forms.

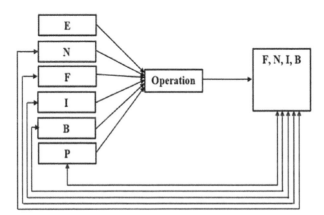

Figure 12.10 Simulation of synergy in variable scenarios

The input of this system is plants (P), environment (E) and different life forms (fungi (F), nematodes (N), insects (I), and bacteria (B)) present in the plant ecosystem. The interaction between these members is antagonism, symbiosis and commensalism. The output of the system is life forms in increase, decrease or intact. This system is a dynamic system, as the double arrows show feedback and feed forward happening between classes of interactions and between different organisms.

Measurement and specific resolution of memberships enable formation of quantifiable systems (Takagi-Sugeno-Kang systems) (Takagi and Sugeno, 1985; Sugeno and Kang, 1988). The latter enable simulation of synergy between and within membership components. TSK systems can be expanded through further mathematic approaches with 100% certainty, creating a function for prediction of inhibition in variable scenarios. Thus, preliminary results presented in this article must be further expanded to determine the details for precise calculation of the interaction via the Takugi-Sugeno-Kang model structure.

Conclusion

This study validates antifungal activity of essential oils of two Moroccan competitive-ruderal plants against the gray mold disease in vitro. Examination with various concentrations of both oils exhibited promising prospects for their utilization against the fungi tested. However, further work is required to indicate the difference in the mechanism of action between the essential oils. This mechanism may be concentration dependent or involve additional metabolites. Simulation of the interaction between plants and different life forms guides the design of plant-based chemical formulations that can promote crop productivity, maintain ecosystem structure and thus maintain and increase diversity within the ecosystem.

References

Banani, H., Olivieri, L., Santoro, K., Garibaldi, A., Gullino, M. L. and Spadaro, D. 2018. "Thyme and savory essential oil efficacy and induction of resistance against *Botrytis cinerea* through priming of defense responses in apple." *Foods* 7 (2):11.

Bicchi, C., Rubiolo, P., Ballero, M., Sanna, C., Matteodo, M., Esposito, F., Zinzula, L. and Tramontano, E. 2009. "HIV-1-inhibiting activity of the essential oil of *Ridolfia segetum* and *Oenanthe crocata*." *Planta medica* 75 (12):1331–1335.

Cabral, C., Poças, J., Gonçalves, M. J., Cavaleiro, C., Cruz, M. T. and Salgueiro, L. 2015. "*Ridolfia segetum* (L.) Moris (Apiaceae) from Portugal: A source of safe antioxidant and anti-inflammatory essential oil." *Industrial Crops and Products* 65:56–61.

Chebli, B., Achouri, M., Hassani, L.M.I. and Hmamouchi, M. 2003. "Chemical composition and antifungal activity of essential oils of seven Moroccan Labiatae against *Botrytis cinerea* Pers: Fr." *Journal of Ethnopharmacology* 89 (1):165–169.

Chebli, B., Hmamouchi, M., Achouri, M. and Hassani, L.M.I. 2004. "Composition and in vitro fungitoxic activity of 19 essential oils against two post-harvest pathogens." *Journal of Essential Oil Research* 16 (5):507–511.

Cid-Pérez, T. S., Torres-Muñoz, J. V., Nevárez-Moorillón, G. V., Palou, E. and López-Malo, A. 2016. "Chemical characterization and antifungal activity of *Poliomintha longiflora* Mexican oregano." *Journal of Essential oil Research* 28 (2):157–165.

DeAlmeida, R.R.P., Souto, R.N.P., Bastos, C.N., Da Silva, M.H.L. and Maia, J.G.S. 2009. "Chemical Variation in *Piper aduncum* and Biological Properties of Its Dillapiole-Rich Essential Oil." *Chemistry & Biodiversity* 6 (9):1427–1434.

Elad, Y., Yunis, H. and Katan, T. 1992. "Multiple fungicide resistance to benzimidazoles, dicarboximides and diethofencarb in field isolates of *Botrytis cinerea* in Israel." *Plant Pathology* 41 (1):41–46.

Fillinger, S. and Elad, Y. 2016. *Botrytis: The Fungus, the Pathogen and Its Management in Agricultural Systems*. Springer International Publishing, Switzerland.

Fleisher, Z. and Fleisher, A. 1996. "Volatiles of leaves and flowers of *Ridolfia segetum* L. Moris. Aromatic plants of the holy land and the Sinai. Part XII." *Journal of Essential Oil Research* 8 (2):189–191.

Fraternale, D., Flamini, G. and Ricci, D. 2016. "Essential oil composition of *Angelica archangelica* L. (Apiaceae) roots and its antifungal activity against plant pathogenic fungi." *Plant Biosystems-An International Journal Dealing with All Aspects of Plant Biology* 150 (3):558–563.

Furze, J.N. 2014. *Global Plant Characterisation and Distribution with Evolution and Climate*, University of the West of England.

Furze, J.N., Swing, K., Gupta, A.K., McClatchey, R.H. and Reynolds, D.M. 2017. *Mathematical Advances Towards Sustainable Environmental Systems*. Springer International Publishing, Switzerland.

Furze, J.N., Zhu, Q., Qiao, F. and Hill, J. 2013a. "Mathematical methods to quantify and characterise the primary elements of trophic systems." *International Journal of Computer Applications in Technology* 47 (4):314–325.

Furze, J.N., Zhu, Q.M., Qiao, F. and Hill, J. 2013b. "Linking and implementation of fuzzy logic control to ordinate plant strategies." *International Journal of Modelling, Identification and Control* 19 (4):333–342.

Grime, J.P. 2006. *Plant Strategies, Vegetation Processes, and Ecosystem Properties*: John Wiley & Sons, New York.

Hammer, K.A., Carson, C.F. and Riley, T.V. 2003. "Antifungal activity of the components of *Melaleuca alternifolia* (tea tree) oil." *Journal of Applied Microbiology* 95 (4):853–860.

Isman, M.B. 2000. "Plant essential oils for pest and disease management." *Crop Protection* 19 (8–10):603–608.

Jabrane, A., Ben Jannet, H., Mastouri, M., Mighri, Z. and Casanova, J. 2010. "Chemical composition and in vitro evaluation of antioxidant and antibacterial activities of the root oil of *Ridolfia segetum* (L.) Moris from Tunisia." *Natural Product Research* 24 (6):491–499.

Jannet, H.B. and Mighri, Z. 2007. "Hydrodistillation kinetic and antibacterial effect studies of the flower essential oil from the Tunisian *Ridolfia segetum* (L.)." *Journal of Essential Oil Research* 19 (3):258–261.

Kordali, S., Cakir, A., Ozer, H., Cakmakci, R., Kesdek, M. and Mete, E. 2008. "Antifungal, phytotoxic and insecticidal properties of essential oil isolated from Turkish *Origanum acutidens* and its three components, carvacrol, thymol and p-cymene." *Bioresource Technology* 99 (18):8788–8795.

Lin, C.C. and Segel, L.A. 1974. *Mathematics Applied to Deterministic Problems in the Natural Sciences*. Macmillan Publishing Co. Inc, New York Google Scholar.

Lughofer, E. 2011. *Evolving Fuzzy Systems-Methodologies, Advanced Concepts and Applications*. Vol. 53. Springer, Berlin.

Mamdani, E.H. 1974. "Application of fuzzy algorithms for control of simple dynamic plant." *Proceedings of the Institution of Electrical Engineers* 121 (12):1585–1588.

Mamdani, E. H. and Assilian, S. 1975. "An experiment in linguistic synthesis with a fuzzy logic controller." *International Journal of Man-machine Studies* 7 (1):1–13.

Marei, G. and Abdelgaleil, S. 2018. "Antifungal potential and biochemical effects of monoterpenes and phenylpropenes on plant pathogenic fungi." *Plant Protection Science* 54 (1).

Marongiu, B., Piras, A., Porcedda, S., Tuveri, E. and Maxia, A. 2007. "Comparative analysis of the oil and supercritical CO2 extract of *Ridolfia segetum* (L.) Moris." *Natural Product Research* 21 (5):412–417.

Mendel, J. M. 2001. *Uncertain rule-based fuzzy logic systems: introduction and new directions.* Prentice Hall, Upper Saddle River, NJ.

Nabigol, A., & Morshedi, H. (2011). Evaluation of the antifungal activity of the Iranian thyme essential oils on the postharvest pathogens of strawberry fruits. *African Journal of Biotechnology,* 10(48), 9864–9869.

Norman, C. (1988). EPA sets new policy on pesticide cancer risks. *Science,* 242(4877), 366–368.

Ouadi, Y. El, Manssouri, M., Bouyanzer, A., Majidi, L., Bendaif, H., Elmsellem, H., Shariati, M. A., Melhaoui, A. and Hammouti, B. 2017. "Essential oil composition and antifungal activity of *Melissa officinalis* originating from north-east Morocco, against postharvest phytopathogenic fungi in apples." *Microbial pathogenesis* 107:321–326.

Palá-Paúl, J., Velasco-Negueruela, A., Pérez-Alonso, M. J. and Vallejo, M.C.G. 2005. "Volatile constituents of *Ridolfia segetum* (L.) Moris gathered in Central Spain: Castilla la Mancha Province." *Journal of Essential Oil Research* 17 (2):119–121.

Palá-Paúl, J., Velasco-Negueruela, A., Pérez-Alonso, M. J. and Ramos-Vázquez, P. 2002. "Volatile constituents of *Ridolfia segetum* (L.) Moris gathered in southern Spain, Andalucia Province." *Journal of Essential Oil Research* 14 (3):206–209.

Pandey, D. K., Tripathi, N. N., Tripathi, R. D. and Dixit, S. N. 1982. "Fungitoxic and phytotoxic properties of the essential oil of *Hyptis suaveolens* [Fungitoxische und phytotoxische Eigenschaften des ätherischen Öis von Hyptis suaveolens]." *Zeitschrift für Pflanzenkrankheiten und Pflanzenschutz/Journal of Plant Diseases and Protection* 344–349.

Pooter, H. L. De, De Buyck, L. F., Schamp, N. M., Aboutabl, E., De Bruyn, A. and Husain, S. Z. 1986. "The volatile fraction of *Senecio glaucus* subsp. *coronopifolius.*" *Flavour and Fragrance Journal* 1 (4–5):159–163.

Rhayour, K., Bouchikhi, T., Tantaoui-Elaraki, A., Sendide, K. and Remmal, A. 2003. "The mechanism of bactericidal action of oregano and clove essential oils and of their phenolic major components on *Escherichia coli* and *Bacillus subtilis.*" *Journal of Essential Oil Research* 5 (5):356–362.

Sekine, T., Sugano, M., Majid, A. and Fujii, Y. 2007. "Antifungal effects of volatile compounds from black zira (*Bunium persicum*) and other spices and herbs." *Journal of Chemical Ecology* 33 (11):2123–2132.

Sharma, N. and Tripathi, A. 2008. "Effects of *Citrus sinensis* (L.) Osbeck epicarp essential oil on growth and morphogenesis of *Aspergillus niger* (L.) Van Tieghem." *Microbiological Research*163 (3):337–344.

Silva, D. T., Bianchini, N. H., Muniz, M.F.B., Heinzmann, B. M. and Labidi, J. 2016. "Chemical composition and inhibitory effects of *Vectandra grandiflora* leaves essential oil against wood decay fungi." *Drewno: prace naukowe, doniesienia, komunikaty* 59.

Soylu, E. M., Kurt, Ş. and Soylu, S. 2010. "In vitro and in vivo antifungal activities of the essential oils of various plants against tomato grey mould disease agent *Botrytis cinerea.*" *International Journal of Food Microbiology* 143 (3):183–189.

Sugeno, M. and Kang, G.T. 1988. "Structure identification of fuzzy model." *Fuzzy Sets and Systems* 28 (1):15–33.

Sugeno, M. and Yasukawa, T. 1993. "A fuzzy-logic-based approach to qualitative modeling." *IEEE Transactions on Fuzzy Systems* 1 (1):7.

Takagi, T. and Sugeno, M. 1985. "Fuzzy identification of systems and its applications to modeling and control." *IEEE transactions on systems, man, and cybernetics* (1):116–132.

Valente, V.M.M., Jham, G.N., Jardim, C.M., Dhingra, O.D. and Ghiviriga, I. 2014. "Major antifungals in nutmeg essential oil against *Aspergillus flavus* and *A. ochraceus.*" *Journal of Food Research* 4 (1):51.

Vizcaíno-Páez, S., Pineda, R., García, C., Gil, J. and Durango, D. 2016. "Metabolism and antifungal activity of safrole, dillapiole, and derivatives against *Botryodiplodia theobromae* and *Colletotrichum acutatum.*" *Boletín Latinoamericano y del Caribe de Plantas Medicinales y Aromáticas* 15 (1).

Wang, L.X. 1997. *A Course in Fuzzy Systems and Control.* Prentice-Hall PTR, Englewood Cliffs, NJ.

Williamson, B., Tudzynski, B., Tudzynski, P. and Van Kan, J.A. 2007. "*Botrytis cinerea*: the cause of grey mould disease." *Molecular Plant Pathology* 8 (5):561–580.

Xueuan, R., Dandan, S., Zhuo, L. and Qingjun, K. 2018. "Effect of mint oil against *Botrytis cinerea* on table grapes and its possible mechanism of action." *European Journal of Plant Pathology* 151 (2):321–328.

Yu, D., Wang, J., Shao, X., Xu, F. and Wang, H. 2015. "Antifungal modes of action of tea tree oil and its two characteristic components against *Botrytis cinerea.*" *Journal of Applied Microbiology* 119 (5):1253–1262.

Zadeh, L.A. 1968. "Communication-fuzzy algorithms." *Information and Control* 12 (2): 94–102. doi: 10.1016/S0019-9958(68)90211-8

Zadeh, L.A. 1973. "Outline of a new approach to the analysis of complex systems and decision processes." *IEEE Transactions on Systems, Man, and Cybernetics* (1):28–44.

Zhang, J.H., Sun, H.L., Chen, S.Y., Zeng, L. and Wang, T.T. 2017. "Anti-fungal activity, mechanism studies on α-phellandrene and nonanal against *Penicillium cyclopium.*" *Botanical Studies* 58 (1):13.

13 Application of project scheduling in production process for paddy cleaning machine by using PERT and CPM techniques

Case study

S. Bangphan, P. Bangphan and S. Phanphet

Introduction

The machine used to mill the rice

The rice mill process is a process of shelling out the paddy to obtain the rice for consumption. The rice mill process may be simple, such as a mortar, or may use modern machinery to mill the rice. The use of a rice mill machine design needs to be appropriate. The quantity of milled rice and the quality of the rice depends on the type and type of machinery used. Good mill performance will include:

- The highest yield of edible rice
- High-quality rice (obtain the best possible quality)
- Minimize losses
- Low cost of rice (minimize the processing cost) in Asian countries.

Recognizing the need for high-quality machinery in the rice mill process, there are also many variables that are related to the quality of rice, such as the moisture content of paddy and the quality of paddy in the harvest. Harvesting should start when the rice moisture content is about 20% and in the dehumidification process. Should be appropriate. Moisture content in paddy storage should be in the range of 13%–14% (wet standards).

Cleaning and sorting (paddy cleaner and sorting) is usually done by farmers at the mill. There are other contaminants such as soil, stone, straw, rice husks, weeds and so forth which need to be cleaned before going into the shell cracking process. This will improve the quality of the cracked rice. Lightweight additives can be separated using a cleaning fan. Weed seeds or other impurities cannot be separated at this stage, and will be separated in the next step. Cleaning machines include the double-sieve cleaner, self-cleaning sieve, single-action aspirator, two-stroke aspirator, double-action aspirator, double-drum pre-cleaner, double-drum pre-cleaner, single-drum pre-cleaner and magnetic separator.

However, there may be other additives or contamination. As a result, the volume of rice has decreased. By operation of the machine, the fan separates the lightweight materials such as the chaff. The medium and large impurities are separated by a cylindrical grating. This system was developed in Japan. The shape of the cylinder screen can be divided into several types, such as a rectangular, round, oblong and triangular. There will be restrictions on cleaning. One cannot clean the paddy 100%, so pre-cleaning is necessary before you put the pads into the machine. Otherwise, clogging may occur.

The sieve is divided into two types: rotating screen and oscillating screen. The paddle moves in the direction of the vibration of the sieve. Therefore, the sieve design is not correct and reduces the performance of the machine.

There will be special equipment for the cleaning machine, for the separation of impurities of the same size as the paddy but with different weights. The principle is that the wind is flowing through the bottom of the rack. With small holes, heavy weights (rock) will be lowered, while the paddy weight is less and moves upwards along the tilt of the sieve. So the rubble is separated from the paddy.

For the separation of iron impurities, the magnetic cleaning mesh is attached. Then scrap metal is removed from the machine periodically by hand.

Sorting refers to the removal of specific materials. Leave the material with different characteristics, for example, to clean the paddy. This will separate the impurities such as straw from the paddy. This material is contaminants, while most paddy or materials are called purity. Features related to cleaning and sorting of grain include:

- Size
- Shape or geometry of product
- Density
- Surface texture
- Mechanical properties of materials
- Electrical characteristics of materials (electrical).

Agriculture is the largest single industry in the world, and seed production is an important segment of this industry. Seed, as it comes from the field, contains various contaminants like weed seeds, other crop seeds, and such inert material as stems, leaves, broken seed, and dirt. These contaminants must be removed, and the cleaned seed properly handled and stored to provide a high-quality planting seed that will increase farm production and supply uniform raw material for industry. The procedures used to meet present quality standards result in a loss of up to 50% of the good seed, even though many special machines and techniques are used for seed cleaning and handling (Harmond et al., 1968). The portable paddy cleaning machine is designed to remove foreign materials and impurities such as sand particles, stones, paddy straws and foreign seeds from the paddy. This machine provides farmers an alternative to the current conventional method should the farmers want to extract the paddy seed in small amounts (Caro Jang et al., 2014). Currently, they only use a traditional winnowing technique as to obtain the seeds to be used next season or before processing paddies to become rice. The performance of this machine is very efficient where the percentage of clean paddy is observed to be at 95%. It helps farmers improvise their traditional method, reduces the purchasing cost of paddy seed and utilizes the cleaning process at low cost and less maintenance (Shayfullzamree et al., 2010). Generally, the hand threshing and the traditional handling used in most developing countries yields a larger percentage of foreign matter with the paddy, thus more cleaning is required. At this point a rice mill cleaner removes any remaining foreign material that could damage the milling machinery and eliminates foreign material from the milled rice (James and Wimberly, 1983).

In this research, the idea was to build a paddy husker cleaner using an activities network. Activity on the arrow and activities on the node to find the critical path in the build machine.

PERT and CPM

The critical path method (abbreviate CPM) network planning was founded in 1956 (Elmaghraby, 2000; Kelley and Walker, 1959). It is the most common technology in graphics mode to take order with project plan. This technology prompt decision-maker to notice that they should focus their attention on critical path, as activities in this path are usually considered to be the most critical part for one project (Liu and Chen, 2006). Definitely, using the technology could scientifically work out the float of each activity in project, and then identify the critical path, point out the critical activities of the project and measure the importance of each activity, thereby improving decision-making ability and management level of the decision maker.

For measure importance of activity, the importance of the arc needs to be measured in the activity-on-arc representation network, and the importance of the node needs to be measured in the activity-on-node representation network (An et al., 2006). The float is the core of CPM network, and its start gone with the appearance of CPM network planning. The conceptions of total float, safety float, free float, and interference float were proposed by Battersby and Thomas (Elmaghraby, 1977) in 1967 and 1969, respectively, and the conclusion was deduced that the path whose total float is zero is the critical path. The conception of node float (Elmaghraby, 1977; Qi, 1997; Li and Qi, 2006; Li et al., 2007) was proposed by Elmaghraby in 1977, and these floats were analyzed by him [8].

Both CPM and program evaluation and review technique (PERT) are network-based techniques and therefore help in programming and monitoring the progress of the stages involved so that the project is completed by the deadline (Taha, 2008). A building project involves a single non-representative scheme typically undertaken to accomplish a premeditated result within a time bound and financial plan. However, because of the individuality of each project, its outcome can never be predicted with absolute confidence (Ayininuola and Olalusi, 2010). It further assists in allocating resources, such as labor and equipment, and thus helps to make the total cost of the building project a minimum by finding the optimum balance between various costs and time involved (Gueret and Sevaux, 2002).

This study is therefore aimed at exploring the time and cost of various activities involved in a paddy cleaning machine in order to determine the optimal completion time using CPM and PERT techniques.

Materials and method

In scheduling network activity, we need to estimate the time each activity should take when performing it in a regular manner. The projection was in accordance with the leader of the paddy rice cleaning machine project in the Sansai district community. Table 13.1 shows the activities related to the production process of paddy cleaners. Construction activity begins with activity A and ends with activity U and shows the distribution of activity of the project relative to the actual number of days to make each activity complete and the cost impact in terms of the three layers to clean the chaff. The cost is labor (student), based on the assumption that the material is already available for use. As the material becomes available, the reduction in the number of days in a particular activity will be affected by the additional labor costs.

Table 13.1 Description of activities paddy cleaning machine

Activity code	Activity Description	Immediate Predecessor	Estimated Duration (Day)	Normal Cost (Dollar)
A	Frame	–	5	115
B	Install wheel base with wheel flange	A	3	55
C	Install bearing sets and shaft	A	2	125
D	Install shaft combine with bearing and pulleys	A	2	135
E	Assembly kit one	B,C,D	3	120
F	Hopper	E	3	85
G	Control hopper sets	E	2	65
H	Shaft support sets	E	3	35
I	Flow controller	E	2	30
J	Assembly kit two	F,G,H,I	4	85
K	Sieve cleaning sets	J	2	140
L	Made shaft of sieve cleaning	J	3	110
M	Crankshaft	J	3	80
N	Belts	J	2	100
O	Pulleys	J	2	120
P	Assembly kit three	K,L,M,N,O	3	135
Q	Top plate	P	1	120
R	Down plate	P	1	120
S	Adjust of belts	P	2	135
T	Assembly kit four	Q,R,S	4	100
U	All assembly kit and adjust	T	5	110

The start and end of each activity, if there are no delays occurring anywhere in the project, are the *earliest start time* and the *earliest finish time* of the activity. The symbols used are (Scaat, 2009 recommented):

ES = earliest start time for a particular activity
EF = earliest finish time for a particular activity
D = (estimated) duration of the activity

where

$$EF = ES + D \qquad (13.1)$$

The *latest finish time* has the corresponding definition with respect to finishing the activity. In symbols:

LS = latest start time for a particular activity
LF = latest finish time for a particular activity

where

$$LS = LF - D \qquad (13.2)$$

The *slack* for an activity is the difference between its latest finish time and its earliest finish time. In symbols:

$$Slack = LF - E \qquad (13.3)$$

The PERT three-estimate approach

The three estimates to be obtained for each activity are

Most likely estimate (m) = estimate of the most likely value of the duration
Optimistic estimate (o) = estimate of the duration under the most favorable conditions
Pessimistic estimate (p) = estimate of the duration under the most unfavorable conditions.

For most probability distributions such as the beta distribution, essentially the entire distribution lies inside the interval between $\mu - 3\sigma$ and $\mu + 3\sigma$ (for a normal distribution, 99.73% of the distribution lies inside this interval.) Thus, the spread between the smallest and largest elapsed times. Therefore, an approximate formula for σ^2 is (Scaat, 2009 recommented).

$$\sigma^2 = \left(\frac{p - o}{6} \right)^2 \qquad (13.4)$$

Similarly, an approximate formula for μ is

$$\mu = \frac{O + 4m + p}{6} \tag{13.5}$$

The project network needs to convey all this information. Two alternative types of project networks are available for doing this. One type is the *activity-on-arc* (AOA) project network, where each activity is represented by an arc. A node is used to separate an activity (an outgoing arc) from each of its immediate predecessors (an incoming arc). The sequencing of the arcs thereby shows the precedence relationships between the activities.

The second type is the *activity-on-node* (AON) project network, where each activity is represented by a node. The arcs then are used just to show the precedence relationships between the activities. In particular, the node for each activity with immediate predecessors has an arc coming in from each of these predecessors. The original versions of PERT and CPM used AOA project networks, so this was the conventional type for some years. However, AON project networks have some important advantages over AOA project networks for conveying exactly the same information.

The standard deviation of the project duration probability distribution is computed by adding the variances of the critical activities (all of the activities that make up the critical path) and taking the square root of that sum probability computations can now be made using the normal distribution table.

Determine probability that project is completed within specified time (Scaat, 2009 recommended).

$$Z = \frac{X - \mu}{\sigma} \tag{13.6}$$

where $\propto$ = tp = project mean time
σ = project standard mean time
X = (proposed specified time).

Critical path

A critical activity is an activity that has no leeway in determining its start and finish times. According to Gaither and Frazier (2002) the critical path is the longest path network and is a chain of critical activities for the project. If a critical activity runs late, then the entire project will run late (Taha, 2008)

A noncritical activity is an activity that allows some scheduling slack, meaning it can be advanced or delayed (within limits) without affecting the completion time of the project (Taha, 2008)

Slack of activity is the maximum length of time that an activity can be delayed without delaying the entire project. Activities on the critical path have zero slack.

Slack of activity is calculated from four time for each activity: (1) earliest start time, (2) earliest finish time, (3) latest start time and (4) latest finish time (Ritzman and Krajewski, 1999).

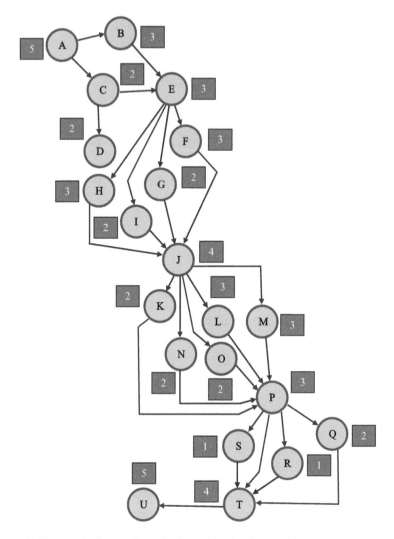

Figure 13.1 Network diagram for activities paddy cleaning machine

Based on the table, we can now construct a network diagram for paddy cleaning machine, shown in Figure 13.1.

Implementation and results

Table 13.2 Activity A starts immediately, so activity value of activity $A = 0$ will calculate the fastest completion time of activity A EF $= 0 + 5 = 5$ days activity U takes 5 days calculate activity LS value $U = 35-5 = 30$ days and calculate the lunar activity. $A =$ LF $=$ EF $= 5 - 5 = 0$. Calculate EF $=$ ES $+ D = 5 + 2 = 7$. The value of J depends on F, G, H and so on. EF $=$ ES $+ D = 14 + 4 = 18$ days.

The value of P depends on K, L, M, N and O. The value of T is from Q, R and S. The maximum value is $R = 30$ (EF), so EF = ES + D = 26 + 4 = 30 days. The LS value of LS = LF is 30 − 4 = 26 days. Activity P is connected to Q, R and S. The minimum value is $R = 24$ days. Therefore, the activity of P = LS = LF = 24 − 3 = 21 days. The remaining values for ES, EF, LS, LF and slack show. Table 13.2 is the projected time of activity and variance of this project.

Critical path

How long should the project last? Summing up, the total duration of the activity will be 31 days. However, this is not an answer to the question, as some activities can be done (approximately) at the same time.

The associated thing is the length of each route through the network.

So should these lengths be the project duration (estimated) (total time needed for the project) reason we come out.

Since the activities on the given paths must be made in sequence without overlapping, the duration of the project must not be shorter than the path length. However, the duration of the project may be longer because some activities on earlier versions of the route may have to wait longer for the previous version to not be completed on the route. For example, consider the second route in Table 13.2 and

Table 13.2 The earliest start, earliest finish, latest start, latest finish times and slack activities of a construction project in paddy cleaning machine

Activity	Predecessors	Duration	the scheduled start and finish times				
			ES	EF	LS	LF	SL
A	0	5	0	5	0	5	0
B	A	3	5	8	5	8	0
C	A	2	5	7	6	8	1
D	A	2	5	7	6	8	1
E	B,C,D	3	8	11	8	11	0
F	E	3	11	14	11	14	0
G	E	2	11	13	12	14	1
H	E	2	11	13	12	14	1
I	E	2	11	13	12	14	1
J	F,G,H,I	4	14	18	14	18	0
K	J	2	18	20	19	21	1
L	J	3	18	21	18	21	0
M	J	3	18	21	18	21	0
N	J	2	18	20	19	21	1
O	J	2	18	20	19	21	1
P	K,L,M,N,O	3	21	24	21	24	0
Q	P	1	24	25	25	26	1
R	P	1	24	25	25	26	1
S	p	2	24	26	24	26	0
T	Q,R,S	4	26	30	26	30	0
U	T	5	30	35	30	35	0

focus on activities E, J, P and T. This activity has two previous generations. One (activity J) is not on the route and one (activity E). E is completed. It takes another three days for Activity J, but it takes four days for activity T and activity U to complete. Therefore, the duration of the project must be longer than the length of the second runway in the table.

Table 13.2 shows the slack for each activity. Note that some activities have zero slack, indicating that delays in these activities will delay the project.

According to Scaat, 2009 that this is how PERT/CPM identifies the critical path. Each activity with a slack center is on an important path through the project network, such that delays along this path will cause the project to delay (Scaat, 2009).

Thus, the critical path is

$$A \rightarrow B \rightarrow E \rightarrow F \rightarrow J \rightarrow L \rightarrow M \rightarrow P \rightarrow S \rightarrow T \rightarrow U$$
$$= 5 \rightarrow 3 \rightarrow 3 \rightarrow 3 \rightarrow 4 \rightarrow 3 \rightarrow 3 \rightarrow 2 \rightarrow \rightarrow 4 \rightarrow 5 = 31 \text{ days}$$

For this reason, the AON network has become increasingly popular with workers. It seems to become a common form used. Therefore, we now will focus only on the AON project network and will reduce the adjective AON.

Figure 13.1 shows the project network for paddy cleaning machine. Table 13.2 shows the first activity until the last activity (A to U) and calculates the values used to construct the machine. Figure 13.2 shows diagram node for AON.

Figure 13.3 shows the AON project network. After placing the activity box in the same position as the corresponding node in Figure 13.3 (except now, the box is not included for the start and end of the project).

Estimating the expected activity time

Consider, for example, the U activity creates all assembly frames and modifies. Construction projects and the following.

Estimates: $o = 5, m = 3$ and $p = 6$

ES	Activity	EF
SL	Description	
LS	Duration	LF

Figure 13.2 Diagram node

Note that each box provides considerable information about the activity. The name is given in the first row. The second row shows the activity number and duration. The last row then gives the scheduled start and finish times.

A relatively large value for A, U is due to delays in the frame and the all assembly kit and adjust. If this unit is delayed, all activities will be delayed.

Activity A and activity U will take quite a long time because the activities begin at the beginning and the activity is near completion. Before proceeding to the next trial.

The responses are shown in the first four columns of Table 13.3. The project network with the duration of each activity is equal to the pessimistic estimate specified in columns 5, 6 and 7 of Table 13.3.

$$\sigma_p = \sqrt{\begin{array}{l} \text{var}(A) + \text{var}(B) + \text{var}(E) + \text{var}(F) + \text{var}(J) + \text{var}(L) + \text{var}(M) \\ + \text{var}(P) + \text{var}(S) + \text{var}(T) + \text{var}(U) \end{array}}$$

$$\sigma_p = \sqrt{\begin{array}{l} 0.028 + 0.028 + 0.028 + 0.111 + 0.028 + 0.028 \\ + 0.028 + 0.028 + 0.028 + 0.111 + 0.028 \end{array}}$$

$$\sigma_p = \sqrt{0.474} = 0.688$$

The fact that activity time is a random variable means that the completion time of a project is a random variable. That is, there is a variance that may occur at the time of complete completion. Although the project is scheduled to be completed within 31 days, there is no guarantee that it will be completed within 31 days. If

Table 13.3 Expected value and variance of the duration of each activity for paddy cleaning project

Activity	o	m	p	(o + 4m + p)/(6)	(p − o)/(6)	σ^2
A	5	4	6	4.500	0.167	0.028
B	3	2	4	2.500	0.167	0.028
C	2	1	3	1.500	0.167	0.028
D	2	1	2	1.333	0.000	0.000
E	3	1	4	1.833	0.167	0.028
F	3	1	5	2.00	0.333	0.111
G	2	1	3	1.500	0.167	0.028
H	2	1	3	1.500	0.167	0.028
I	2	1	3	1.500	0.167	0.028
J	4	2	5	2.833	0.167	0.028
K	2	1	3	1.500	0.167	0.028
L	3	1	4	1.833	0.167	0.028
M	3	1	3	1.500	0.000	0.028
N	2	1	3	1.667	0.167	0.028
O	2	2	4	2.333	0.333	0.111
P	3	3	4	3.167	0.167	0.028
Q	1	1	3	1.333	0.333	0.111
R	1	1	3	1.333	0.333	0.111
S	2	1	3	1.500	0.167	0.028
T	4	3	5	3.500	0.167	0.111
U	5	3	6	3.833	0.167	0.028

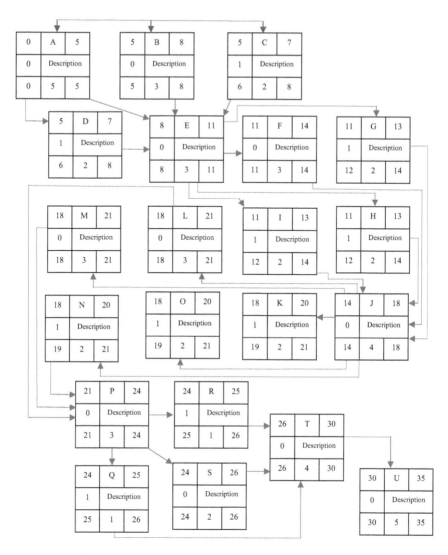

Figure 13.3 Diagram node for constructs of paddy cleaning machine

the activity exceeds the expected time, the project may not be completed within 35 days. Date required in general, it is useful to know the feasibility of a project to be completed within a specified time. Particularly, cleaning paddy needs to be aware of the possibility of moving within 35 days.

So that:

$$Z = \frac{X - \mu}{\sigma} = \frac{35 - 31}{0.688} = 5.81$$

The probability that the project will be completed in 35 days is equivalent to 99.999%. The probability that the project will be completed is shown in Figure 13.4.

Considering time-cost trade-offs

This research must consider how much additional cost reduction will be required to reduce the projected period to 28 days (the research deadline for the bonus is $50 for completion) will answer the next question at the end of Table 13.3.

The first key concept for this approach is that of *crashing*.

Crashing an activity refers to taking special costly measures to reduce the duration of an activity below its normal value.

The cost-benefit (CPM) approach of time-cost trade-offs involves determining how much (if any) each activity is faulty to reduce the projected time-frame. Project to get the desired value.

The information needed to determine the number of crashes in any activity is given by the activity graph. Figure 13.5 shows the typical cost graph. Notice

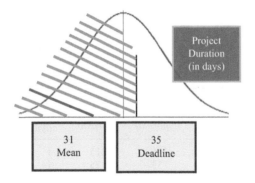

Figure 13.4 The shaded area is the portion of the distribution that meets the deadline of 35 days

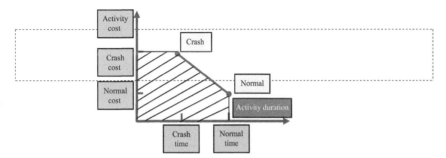

Figure 13.5 A typical time-cost graph for an activity

two important points on this graph: normal and crash (Scaat, 2009; Surapong, 2013).

The student research staff responsible for arranging the entire assembly and modifying the two temporary staff additions and overtime hours will allow him to reduce the duration of this activity from five days to three days, which is possible. The research staff then assessed the full cost of doing so in comparison to the five-day schedule as shown below.

Activity U (put up the all assembly kit and adjust):

Normal point: time = 5 days, cost = $110.

Crash point: time = 3 days, cost = $300.

Maximum reduction in time = 5 − 3 = 2 days.

$$\text{Crash cost per day saved} = \frac{\$300 - \$110}{2} = \$95$$

Table 13.4 gives the corresponding data obtained for all the activities. Summing the *normal cost* and *crash cost* columns of Table 13.4 gives

Sum of normal costs = $2,120,

Sum of crash costs   = $4,775.

Table 13.4 Time-cost trade-off data for the activities of paddy cleaning project

Act. code	time		cost		Maximum reduction in time	Crash cost per day saved
	normal	crash	normal	crash		
A	5	4	115	215	1	100
B	3	2	55	175	1	120
C	2	1	125	250	1	125
D	2	1	135	275	1	140
E	3	1	120	380	2	130
F	3	2	85	195	1	110
G	2	1	65	190	1	125
H	3	1	35	235	2	100
I	2	1	30	125	1	95
J	4	2	85	295	2	105
K	2	1	140	225	1	85
L	3	1	110	345	2	118
M	3	1	80	295	2	108
N	2	1	100	225	1	125
O	2	1	120	235	1	115
P	3	2	135	245	1	135
Q	1	0	120	0	0	0
R	1	0	120	10	0	0
S	2	1	135	275	1	70
T	4	2	100	295	2	98
U	5	3	110	300	2	95

This payment is required to cover some overhead costs in addition to the costs of the activities listed in the table, and to provide reasonable returns to the research staff. When developing a $2,500 bid, trust management felt that this amount would provide a reasonable profit as long as the total cost of the event was fair, close to the normal $2,120. Research staff responsible for making the project as close to the budget and as possible.

As has been seen in Figure 13.1, if all activities are carried out in a normal manner, the expected duration of the project is 31 days (if delays can be avoided).

If all events are completely crashed, instead of the same calculation, it will be reduced to 28 days, but look at the forbidden costs ($4,775).

Fully crashing all activities clearly is not an option that can be considered. Longest length (35 days) the only way to reduce project duration by day is to shorten the duration of activities on this specific route within one day.

Compare the cost per day recorded in the last column of Table 13.4. For these activities, the smallest cost was $75 for activity S, therefore the first change was to reduce activity S to reduce the time by day.

Conclusion

Table 13.2 shows the expected and variance of the duration of each activity shown in Table 13.3 and shown in Figure 13.3. The components and details are very high. Table 13.2 shows the activities.

As shown in Table 13.1 and Table 13.2, each activity must be merged. Some activities have been skipped. The production plan is not in line with the project plan, that is, the machinery and equipment used in production.

There are many users. The machine and equipment belong to the university. In this study, the maximum period of 35 days was 31 days. The major critical path or floating time was 31 days.

The project management was based on a network of activities to control the production process of paddy cleaning machines.

As to plan performance, there are steps in doing each activity. (Some of the time periods of each activity do not correspond to the plan.)

Each activity is linked to all other activities. It takes less time to create the machine. The project can work for 35 days and the cost of activities that may be held fairly is close to the normal level of approximately $2,120.

The possibility of the project being completed within 35 days is 99.999%; 0.688 is needed for experienced students, and expertise in the use of machinery and equipment in this project. Using the network to analyze waste reduction (less waste) takes less time to produce, because this research controls all production processes using the above activity network. The quality of rice cleaning is equivalent to commercial production.

Acknowledgments

Financial support "RMUTL Research grants for agricultural research, such as cutting and packaging, cold storage, processing and packaging. Research and

202   *S. Bangphan, P. Bangphan and S. Phanphet*

development of economic crops. Project Code AG2–180307163528–60": Rajamangala University of Technology Lanna, Chiang Mai is gratefully acknowledged.

References

An, S.H., Jie, P.R. and He, G.G. 2006. Comprehensive Importance Measurement for Nodes within a Node-Weighted Network, *Journal of Management Sciences in China*, 9 (6): 37–42.

Ayininuola, G.M. and Olalusi, O.O, 2010. Assessment of Building Failures in Nigeria: Lagos and Ibadan Case Study, *African Journal of Science and Technology (AJST). Science and Engineering Series*, 5 (1): 73–78.

Bangphan, Surapong. 2013. *Project Management Engineering, Teaching Materials, Translation and Collection*. Chiang Mai. Binding to cover by RMUTL.

Elmaghraby, S.E. 1977. *Activity Networks: Project Planning and Control by Network Models*, New York: John Wiley & Sons Inc.

Elmaghraby, S.E. 2000. On Criticality and Sensitivity in Activity Networks, *European Journal of Operation Research*, 127: 220–238.

Gaither, N. and Frazier, G., 2002. *Administração da Produção e Operações*. 8ª. Edição. São Paulo: Thomson Learning.

Gueret, C.C. and Sevaux, M., 2002. *Applications of Optimization with Xpress- MP, translated and revised by Susanne Heipke*, Dash Optimization Ltd., London. First Edition, pp. 159–178.

Harmond, Jesse, E., N., Brandenburg, Robert and Klein, Leonard M. 1968. *Agriculture Handbook No. 179, Seed Cleaning and Handling. Agriculture Handbook No. 354, Agricultural Research Service U.S. Department of Agriculture in Cooperation with Oregon Agricultural Experiment Station.*

James, E. and Wimberly, 1983. *Technical Handbook for the Paddy Rice Postharvest Industry in Developing Countries*, International Rice Research Institute. Los Banos, Laguna, Philippines.

Jang, Caro, Kahn, Nada, Rodgerliu, Leandralanglois, and Montanaro, Gianni. 2014. *Design of a Winnowing Machine for West African Rice Farmers*. Department of Bioresource Engineering McGill University.

Kelley, J.E. and Walker, M.R. 1959. Critical Path Planning and Scheduling. In *Proceedings of the Eastern Joint Computational Conference*, 16: 160–172.

Li, X.M. and Qi, J.X. 2006. Study on Time-Scaled Network Diagram Based on the Analysis of Activity Floats, *Operations Research and Management Science*, 15 (6): 28–33.

Li, X.M., Qi, J.X. and Su, Z.X. 2007. The Research Resource Leveling Based on the Analysis of Activity Floats, *Chinese Journal of Management Science*, 15 (1): 47–54.

Liu, C.L. and Chen, H.Y. 2006. Critical Path for an Interval Project Networks, *Journal of Management Sciences in China*, 9 (1): 27–32.

Ritzman, L.J. and Krajewski, L.P., 1999. *Operations Management: Strategy and Analysis*, 5th Edition. USA: Addison-Wesley Publishing Company.

Scaat. [Internet], 2009. [cited 2013 July 12]. www.scaat.in.th/Bachelor/new/1_2552/010.

Shayfullzamree, Rahim, Mohdfathullah, Ghazli, Mohdazaman and Mdderos, 2010. Design of a portable paddy cleaning machine, *Proceedings of the International Conference on Design and Concurrent Engineering*. http://hdl.handle.net/123456789/14018.

Taha, H.A., 2008. *Pesquisa Operacional*. 8a Edição. São Paulo: Pearson Prentice Hall.

Qi, J.X, 1997. *New Theory of Network Planning and Technical Economics Decision-making*, Beijing: Science Press.

14 The management of deep learning algorithms to enhance momentum trading strategies during the time frame to quick detect market of smart money

Khalid Abouloula, Ali Ou-Yassine and Salah-ddine Krit

Introduction

With the news of the stock market and database connection tools such as MetaTrader, with which it is now possible to connect to several stock exchanges like NYSE, NASDAQ, MOEX and Forex without having to open several windows, it's time to search for a secure, improved connection to the stock market, to open the area to all and to simplify trading for companies to save time. All this cannot be achieved unless a long-term vision will be put in place to improve trading strategies and manage deep learning by implementing artificial intelligence (AI), machine learning (ML) on a big data strategy and blockchain network. Switching to artificial intelligence and machine learning on data to improve the ability to sell and buy can better manage risk based on big data analysis [1]. Trading results in a large amount of data, and this is typically desired by machine learning to work effectively [2]. Tokens can be used to represent polling power in an organization, paid credits within an application and almost anything imaginable. Tokens on the blockchain can represent any digital asset; the development of such a blockchain token is an opportunity to achieve communication; the model must be searched for in historical data and the search process should be continuous according to the efficiency of the model currently in use. In practice and at a given moment in the financial time series, a possible model design provides a better plan than any other model in terms of estimated function.

Generally, algorithmic trading is the use of a computer program to create automation in one or several steps of the trading process, depending on the technology used, the objective or where in the trader pipeline the automation occurs. Algorithmic trading includes the following sub-categories [3]: systematic trading (ST), or systems that use trading strategies based on a set of predefined rules received as human input; and high-frequency trading (HTF), which is usually used in systems characterized with fast execution speed (on the order of milliseconds) and holding stock for a short time span. Usually an HFT algorithm proceeds without the intervention of a human; there is an infinite number of trading algorithms, but there are not too many winning solutions, because it brings a lot of money to hedge funds, consulting companies, brokers, traders and so forth.

In the past 20 years the difference between what buyers want to pay and sellers want to be paid has fallen dramatically. One of the reasons for this is the increase in preciseness: stock prices have gone from trading in fractions to pennies; HFT has also added more liquidity to the market, eliminating high bid-ask spreads that once were prevalent. As the study by Elite Group indicates, a lower bid-ask spread helps an average retail US trader save up to $250 every year alone. The idea is to get access to data milliseconds faster than those traders waiting patiently.

Trading platform

The probability and calculations are the basis of trading, and computers make calculations faster. A trading strategy is profitable, it should be repeated often without missing a chance, and emotions should not affect the trading of financial instruments. Trade has turned to more computers very quickly because computers have no emotions and can track thousands of transactions at a time and respond to opportunities in microseconds. Humans are affected by greed and fear, and the human eye can only monitor 10–15 stocks at a time [4].

Principal of the automatic

The purpose of the automatic is to manufacture machines where the human does not intervene except to give orders, or instructions, which are called regulators capable of changing the behavior in the sense that we want the systems considered [5]. An automatic is a set of scientific disciplines and techniques used for the design and use of devices that operate without the intervention of a human operator such as scientific disciplines: this suggests that the automatic requires some theoretical activities in order to achieve: a mathematical modeling of a device, an analysis of its properties on the basis of the model and the design of a control law always based on the model, and design and use of devices. This is implementation that can involve disciplines such as electronics, computers and so forth.

Without the intervention of a human operator – this last expression brings up the notion of automated systems that improve the performance and safety of a device. In automatic, the notion of system is unavoidable. The definition given by the automation specialist is similar to the classical one borrowed from physics [6]. Generally, the system is a device that works in interaction with its environment, generating a set of phenomena. Some physical quantities of the environment act on the system, called inputs; others emanate from the system and act on the outside, called outputs.

A system is described by the differential equation:

$$\dot{y} = u \tag{14.1}$$

The transfer function of this system is the Laplace transform of the output for zero initial conditions and for an input $u(t) = \delta0(t)$.

The Laplace transform of the equation is:

$$s*^\wedge y(s) - y(0) = \hat{u}(s) \tag{14.2}$$

For $y(0) = 0$ and for $\hat{u}(s) = 1$ (that is, for $u(t) = \delta0(t)$), we obtain $y(s) = 1/s$. So the transfer function of the integrator is:

$$H(s) = 1/s \tag{14.3}$$

Its frequency response is:

$$H(iw) = a(w) e^{i p(\omega)} = \frac{1}{iw} \tag{14.4}$$

It is considered a particular first order system for economic and financial systems, for which this formalism may seem less obvious.

Artificial intelligence and machine learning in trading

Artificial intelligence (AI) and machine learning (ML) are being rapidly adopted for a range of applications in the financial services industry. The reliance on the automated learning algorithm is the first destination for e-security companies as it is a haven for dealing with malicious networks. The automated learning algorithm in turn learns about training on massive data and then deals with networks and updates faster. The algorithm of automated learning depends on the e-security companies, and their approach runs counter to the idea of artificial intelligence they call. The approach to automated learning and development of malicious malware and the addition of electronic security companies to sign as a distinctive mark and control the devices of their customers when they appear. An automated learning approach simulation training algorithm for huge catalogs to deal with malicious networks has only increased its flexibility when malware appears. As such, it is important to begin considering the financial stability implications of such uses. Because uses of this technology in finance are in a nascent and rapidly evolving phase, and because data on usage are largely unavailable, any analysis must be necessarily preliminary and developments in this area should be monitored closely.

Artificial intelligence (AI) is a human experience of computational tools to address tasks. As a field, AI has existed for many years. However, recent increases in computing power coupled with increases in the availability and quantity of data have resulted in a resurgence of interest in potential applications of artificial intelligence [7]. These applications are already being used to diagnose diseases, translate languages, and drive cars; they are increasingly being used in the financial sector as well [8].

Experts define that big data is any set of data that is larger than the capacity to be processed using traditional database tools to capture, share, transfer, store, manage and analyze within an acceptable time frame. From the point of view of service providers, organizations need to deal with a large amount of data for the purpose of analysis [9]. Because of the time, effort and cost that large data needs to analyze and process, technicians have to rely on artificial intelligence systems that have the ability to learn, infer and react to situations that have not been programmed into the machine using complex algorithms to work on, as well as using cloud computing techniques to complete their work [10].

Since the financial crisis, much has been done in terms of controlling an online broker not allowing clients to place orders for a trader with an account of 20,000 euros with a limit of ten contracts per day and what should automatically be blocked, when it was not enough because after ten years a new history reveals the deficiencies of the financial markets [11]. A trader, acting on his own, would have taken a cumulative position of 5.6 billion euros. Knowing that, his account open as a broker was limited to 20,000 euros.

Massive data requires exceptional techniques to handle large amounts of data efficiently within the time limit. Massive, multidimensional data can be handled more efficiently by computational calculations. Additional technologies being applied to large data include massive parallel databases, search-based applications, data and mining networks, distributed file systems, distributed databases, cloud-based infrastructure and the internet.

Harmonic patterns

Harmonic models are a trading method in which the trader identifies a pattern of W or M on the graph and derives Fibonacci corrections of the ribs on the graph. There are many harmonic models, and they are different from different levels of Fibonacci corrections and length of ribs. These models depend on the fact that prices move in periodic waves in a gradual manner instead of moving them in a straight and permanent way [12]. In order to apply this model in the Forex world, the trader must know the theory of Fibonacci corrections. The Italian mathematician Fibonacci discovered a series of numbers whose values were present in nature.

Harmonized price patterns take the engineering price patterns to the next level using Fibonacci numbers to determine accurate turning points. Unlike other trading methods, symmetrical trading attempts to predict future movements. This is a broad contrast to the common methods that are retroactive and not predictive. Harmonic trading or harmonic is a precise and mathematical way of trading and trading, but it requires patience, practice and a lot of study to master the patterns and know the forms and engage in transactions accordingly. The basic measurement is only the beginning. Movements that are not in line with the appropriate pattern measurements invalidate the pattern and can lead to misleading the trader so you have to make sure of the form or model before entering the deal depending on it.

Gartley, butterfly, bat and cancer models are the common patterns seen by traders. Entries are made in the potential reversal area when the price confirmation indicates a reversal, and the stop loss is placed under long entry or short entry, or alternatively outside of the form's display.

The harmonious trading of patterns and mathematics combines in a precise trading mode and on the assumption that patterns are duplicated. At the root of the methodology is the initial ratio, or some derived from it (0.618 or 1.618). Supplementary ratios include 0.382, 0.50, 1.41, 2.0, 2.24, 2.618, 3.14 and 3.618. The primary ratio is present in almost all natural and environmental structures and events. They are also found in man-made structures. As the pattern is repeated throughout nature and within society, the ratio is also similar in financial markets that are affected by the environments and communities that trade [13].

Harmonic models are the most accurate set of trends in the trade. The Gartley model, however, is said to be one of the most accurate models among harmonic price models. Precision is the strength of the Gartley model. If we want a system that is accurate, then trading on the Gartley model could be for us. Another advantage of the Gartley model is its tight stop losses. Because of its precise turning points, we have the freedom to set tight stop losses, which contributes to a healthy risk-reward ratio.

Finally, since the Gartley pattern provides for an extension of 1.618 Fibonacci ratios, it gives us the luxury of winning a large amount of glitches, while risking a few on a tight stop loss. Trading on the Gartley model is one of those systems where the odds are in our favor. Pair it with a money management system, and we will be on our way to a steady increase in our account.

Momentum trading strategies

Academic studies have found that the most deliberative strategies that can only be built on historical price data are trading strategy based on time series momentum. This can be done by momentum traders simply by choosing a variety of tradable instruments and buying those that go up and sell down. This method actually tends to make generally larger profits than adding a "best" filter, although there's no reason why you should not use basic analysis or any other filters instead [14].

That the momentum index is within the directional indicators in the trading indicators that follow the main movement of the currency, which is one of the indicators that measure the rate of change and activity in the movements of currency rates, knowing that this indicator shows the trend of market prices for a period of time, both high or low. The entry or exit points of the market may help the investor in his strategy, which is why it is also known as the indicators of strength and payment, because they contain important entry and exit points for the trader, knowing that this indicator is used in particular to determine a specific trading range in order to contribute to the identification market prices and forecast future currency rates [15].

The momentum indicator is rated as an oscillator. It is designed to determine the strength of price action, it helps the trader to assess the extent of price fluctuations over a specified time period, additionally to identify the sensitive conditions that can be exploited to open a deal. The momentum indicator includes one curve that fluctuates within the additional window boundaries. Any extreme deviation emphasized by the technical tool can be used to successfully open the trade. This forms the basis of the momentum strategy [16].

The momentum can be interpreted literally as the volume of motion. It is calculated as a percentage of the current market price to the price that has been installed since a given point in time. The following calculation method assumes that the center line of the oscillator is at the 100 level. The longer the index parameter (time period), the slower the line fluctuates and the less responsive it is to various price changes [17]. We identify the selling and buying signals with the momentum indicator as follows:

- On the upside, the oversold areas are ignored and the oversold points are used to open a long position.
- In the bearish direction, oversold areas are ignored and tap points are used to open a short position.
- In the horizontal direction, take advantage of the areas of sales saturation and oversold areas in the sale.

Momentum indicators are ideally done in case the market moves sideways, but it may not give strong results for the expectation of price reversal points if the market is in a trending position, and that is why it is necessary to pay attention to the trend and consider it when dealing with the index momentum [18].

In Hamiltonian dynamics, the equation of motion of a conservative system is written as:

$$\dot{x} = \frac{\partial H}{\partial p}; \quad \dot{p} = -\frac{\partial H}{\partial x} \tag{14.5}$$

where x is the spatial variable, $\dot{p}$ is the momenta of x and H is the total energy.

Here is a trade station code used in trading, with a simple momentum entry:

```
Input: moment ();
If close>close [moment] then buy next bar at market;
If close<close [moment] then sell next bar at market;
End;
```

Big risk is often commensurate with big bonuses, as equity moves are over-radicalized as sales are short and losses are also stopped. Often as a reaction to incentives, the momentum of profit taking is being traded in equities that are making heavy price gains and falls on heavy volumes.

Tools and environment

Currently, a simple PC or Mac as hardware is perfectly suited to start trading in the markets; it is not necessary to have a very powerful workstation. Trading platforms are software that is not very resource intensive, the most popular is MT4

(Meta Trader 4) and Ninja Trader; an ADSL internet connection will also do the trick [19].

The trading platforms are tied with certain brokerage companies and data vendors. Internationally, small markets such as the Helsinki stock exchange and related brokerage companies are not necessarily supported. However, in order to access historical market data or execute actual live trading, either a license or fees are to be paid. A full trading platform license costs can form a major part of typical retail trader budget. The platforms offer customizable options and programming environment, they still place restrictions and limitations what can be done programmatically [20].

The situation can be look at from a retail investor, a broker company and a stock exchange point of view. Retail investors typically follow price and other information of financial instruments via computer systems, to be a desktop, a laptop or a handheld device. Placing transactions to brokers and further to stock exchanges takes place via same computer systems.

Brokers

Indeed the broker is the intermediary between traders and markets, and without broker it is not possible to trade, the choice of broker is a subject of great concern to traders, the broker is one who will provide a trading platform and will pass through this market flows, traders will see real-time graphs on the evolution of the market second by second and can position to buy or sale from a fraction of a second [21]. Some of the popular brokers for the markets include the following.

Dealing desk

A dealing desk broker also called a "market maker" is a broker who instead of transmitting orders to the interbank market will play the role of market itself. Depending on the history and results in the short and long term it will establish a strategy to cover its risks. The dealing desk or market maker brokers are the most common brokers; they are the easiest to find and they often do aggressive marketing, offer bonuses and often offer fixed spreads. The disadvantage of dealing desk brokers is that there is a conflict of interest between the trader and the dealing desk broker. If the broker applies the strategy to position himself in the opposite direction of trades, when the traders lose, the broker wins.

No dealing desk

A "no dealing desk" broker will transmit orders to the interbank market. They often work with several banks and forward traders' orders to the bank that offers the best price at the time of the transaction. The bank can accept or refuse the order (this is called the "last look"). They are paid on the spread. There are two types of "no dealing desk" brokers:

• STP (straight through processing) that place orders to banks and pay through the spread.

- The ECN (electronic communications network), which centralizes the different market positions between banks, market makers and traders to offer the best possible spread at any time.

Disadvantage of brokers no dealing desk is that brokers offer orders to banks that can refuse (last look), which leads to offer the trade to another bank, extends the lead time and can cause a slippage.

MTF (multilateral trading facility)

These are alternative stock markets created to compete with major historical stock markets. They are a centralized marketplace that allows buyers and sellers to meet. The disadvantage would be a limit of liquidity in the Forex market, which is a very liquid market; the problem does not arise in this market and orders are always executed in a split second.

Spreads

Each broker offers different spreads for parity. A spread is the difference between quotations of supply and demand; it is the commission that brokers collect on your transactions which they directly depend on agreements negotiated between brokers and banks. Naturally, more than the commission is weak, it's better [22].

The spreads are divided into two parts:

- Fixed spreads: These are the most common. This means that for a given parity, the spread is fixed and established in advance. Most brokers offer fixed spreads. Note that they are generally higher than the variable spreads.
- Variable spread: These are present on ECN-type no dealing desk brokers. They vary according to the supply and demand, the volatility and the transaction volume at time t. They are generally weaker than fixed spreads but are only available to experienced traders who trade currencies on "expert" brokers.

Parity

On the Forex, traders cannot buy only dollars or euros; they always buy a currency pair or "parity" Currency1/Currency2 as EUR/USD. Currency 1 is the "currency of reference" and Currency 2 is called "against currency". The most bought and sold exchange rates will have high volatility and will therefore be more attractive to trade, such as EUR/USD, GBP/USD, USD/JPY and USD/CHF. These are "major" parities; they represent a large part of transactions recorded on the Forex.

Pips

This is the unit of variation of Forex prices. On the EUR/USD it is 0.0001, but as most brokers are five digits behind the point, a tenth of a pip will also appear.

Table 14.1 The simile of spreads

	EUFUU SD	USD/ JPY	GBP/ USD	USDICHF	U SINCAD	EURIGBP	GBPIJPY
eToro	3 pips	2 pips	4 pips	3 pips	3 pips	4 pips	6 pips
Plus500	2 pips	2 pips	4 pips	5 pips	5 pips	5 pips	1111bps
EasyForex	3 pips	3 pips	4 pips	3 pips	4 pips	3 pips	8 pips
Forexyard	2.2 pips	2.4 pips	2.8 Dips	2.9 pips	3.6 pips	3.1 pips	5.4 pips
Finexo	3 pips	3 pips	3 pips	3 pips	4 pips	4 pips	7 pips
AvaFX	3 pips	3 pips	4 pips	4 pips	4 pips	3 pips	7 pips
4XP	3 pips	3 pips	4 pips	4 pips	5 pips	0.8 pips	8 pips
FxPro	0.6 pips	0.6 pips	0.9 pips	1.6 pips	1 pips	1.5 pips	1.9 pips
Alpari	1.6 Dips	2 pips	2.5 Dips	2.6 pips	3.5 pips	2 pips	5.5 pips
ActivTrades	0.5 pips	0.5 pips	0.8 pips	1 pips	1 pips	0.8 pips	2.9 pips
ECMarkets	2 pips	3 pips	2 pips	3 pips	3 pips	3.1 pips	5.3 pips
XM.com	1.7 pips	1.8 pips	2.5 pips	3 pips	2 pips	2 pips	4 pips
XTB	0.9 pips	1.8 pips	2.5 pips	1.9 pips	2.2 Dips	2.1 pips	5.5 pips
UFXMarkets	4 pips	**5 pips**	**5 pips**	**5 pips**	**MEM**	-	-
IronFX	1.8 pips	2.2 pips	2.3 pips	2.6 pips	2.7 pips	2 pips	2.7 pips
AvaTrade	3 pips	3 pips	4 pips	4 pips	4 pips	3 pips	7 pips
Trade*	2.5 pips	2.5 pips	2.5 pips	2.8 pips	2.8 pips	2.5 pips	3.5 pips
NESS1x	2.6 pips	2.6 pips	3.5 pips	3.5 pips	4 pips	3 pips	3 pips

So you will see for the EUR/USD 0.00010, which corresponds to one pip and not to ten pips. On yen pairs like EUR/YEN or USD/YEN you will see 0.010 for a variation of one pip.

Deep learning

Deep learning is a branch of machine learning science, a field in which the computer tests logarithms and programs and learns to improve and develop it by itself. The field of machine learning is not new and its roots date back to the middle of the 20th century. The English mathematician Alan Turing proposed a vision of artificial intelligence (the learned machine). Over the next decades, several artificial intelligence techniques have emerged and disappeared. One of these techniques is neural networks. The algorithms that support deep learning play a pivotal role in robotics.

Deep learning is a machine learning method that analyzes data in multiple layers of learning. It may start doing so by learning about simpler concepts and combining these simpler concepts to learn about more complex concepts and abstract notions. It is often said that the principal object of automation is to simplify tasks that will be easy for people to define but are boring to perform [23]. On the other hand, the goal of deep learning AI systems is to perform tasks that are difficult for people to define but easy to perform.

Often referred to as "machine learning" or "neural networks" and "big data", they are closely related to deep learning, which is understood as "training" for machines and computers are an arithmetic model that enables computers or machines to understand natural language. Once computers are equipped with the data and information to understand these relationships, they can come out in a certain sense or perform a task based on logical reasoning, and with the abundance of data and information obtained by the computer or machine, it is supposed to "learn" or gain new experience.

We have an ensemble of N binary spins (Xi) taking values in the set $(1, -1)$ N.

Referring to the Hamiltonian as $H(Xi)$, we assign probability to a spin configuration through the Boltzmann distribution as:

$$P(Xi) = \frac{e^{-H(Xi)}}{Z} \tag{14.6}$$

With the partition function:

$$Z = Tr\, H(Xi) \tag{14.7}$$

The Hamiltonian is characterized by couplings K between the spin interactions equivalent to correlations in deep learning as:

$$H(Xi) = -\sum_i KiXi \sum_{i,j} KijXiXj - \dots. \tag{14.8}$$

In deep learning research the python language is commonly used. There are also extensive libraries on data analysis and machine learning [24].

Results and discussions

Related work

According to studies and analysis done on brokers, we noticed that the majority of brokers that exist in the market and which are used by the platform Meta Trader is based on a single signal broker that follows a single indicator in addition to trailing indicator which rents less reliable and takes more risk to use. These studies often take the approach of trying to find the best algorithmic trading strategy, and for that purpose provide a test bed to evaluate the strategies' performance and compare the results.

It is clear that a difference period between the trend broker and the market signal will be the principal bias of losing for investors.

We propose in this paper a more reliable broker and fix at zero risk level. It is important to time the buys and sells correctly to avoid losses by using proper risk management techniques and stop losses. Implementing a momentum trading index is now very important in trading, because according to the effective market hypothesis and in addition to take very fast decision compared to the other brokers, and using the combination of two compatible signals in function and same criterion of trading, we tell that the signals observed each of them other by choosing a slight difference in period to obtain the two on the same line unlike the first which are off the good predict.

Algorithm: Broker

```
input string Inp_Broker_Title ="BrokerDeM";
int Broker_MagicNumber =102;
bool Broker_EveryTick =false;

input int Inp_SignalDemarker_PeriodFast =8;
input int Inp_SignalDemarker_PeriodSlow =24;
input int Inp_SignalDemarker_PeriodSignal=16;
input int Inp_SignalDemarker_TakeProfit =100;
input int Inp_SignalDemarker_StopLoss =50;

int OnInit(void)
  { if(!ExtBroker.Init(Symbol(),Period(),Broker_
    EveryTick,Broker_MagicNumber))
    {
    printf(__FUNCTION__+": error initializing broker");
    ExtBroker.Deinit();
    return(-1);
    }
  CSignalDeM *signal=new CsignalDeM;
  if(signal==NULL)
    {
    printf(__FUNCTION__+": error creating signal");
    ExtBroker.Deinit();
    return(-2);
    }
  if(!ExtBroker.InitSignal(signal))
    {
    printf(__FUNCTION__+": error initializing signal");
    ExtBroker.Deinit();
    return(-3);
    }
  signal.TakeLevel(Inp_Signal_DeM_TakeProfit);
  signal.StopLevel(Inp_Signal_DeM_StopLoss);
  if(!signal.ValidationSettings())
    {
    printf(__FUNCTION__+": error signal parameters");
    ExtBroker.Deinit();
    return(-4);
    }
  CtrailingFixedPips *trailing=new CtrailingFixedPips;
  if(trailing==NULL)
```

```
    {
    printf(__FUNCTION__+": error creating trailing");
    ExtBroker.Deinit();
    return(-5);
    }
  if(!ExtBroker.InitTrailing(trailing))
    {
    printf(__FUNCTION__+":      error      initializing
    trailing");
    ExtBroker.Deinit();
    return(-6);
    }
  if(!trailing.ValidationSettings())
    {
    // - failed
    printf(__FUNCTION__+": error trailing parameters");
    ExtBroker.Deinit();
    return(-7);
    }
  CmoneyFixedRisk *money=new CmoneyFixedRisk;
  if(money==NULL)
    {
    printf(__FUNCTION__+": error creating money");
    ExtBroker.Deinit();
    return(-8);
    }
  if(!ExtBroker.InitMoney(money))
    {
    printf(__FUNCTION__+": error initializing money");
    ExtBroker.Deinit();
    return(-9);
    }
  if(!money.ValidationSettings())
    {
    printf(__FUNCTION__+": error money parameters");
    ExtBroker.Deinit();
    return(-10);
    }
  if(!ExtBroker.InitIndicators())
    {
    printf(__FUNCTION__+":      error      initializing
    indicators");
    ExtBroker.Deinit();
    return(-11);
    }
  return(INIT_SUCCEEDED);
  }
```

Table 14.2 The windows of result data

Open	1.18874
High	1.19041
Low	1.18801
Close	1.19007
Spread	2

Back testing

Previous test results show profit and loss and some common performance statistics that help determine the return of the risk-adjusted strategy. The test program can be a great addition to automated trading, as the test program uses future information to identify current trading signals.

We get the expected result, and until we control automatically and more accurately at the time of sale and purchase, we work on adapt parameters as shown in figure (6). Is figure 6 after insertion the indicators and changing in the period and shift parameters and apply to typical price in figure 5. To determine when to buy and when to sell, we use momentum trading, as the simulation results give positive momentum to get positive returns in the near future. With high prices, buyers are attracted; with negative momentum there are negative returns; and sellers are attracted by low prices.

It has been shown how an automated method was easily linked to a structure with approximate data, from which the controls were tightly linked through both simple and complex data. Thus, data binding refers to the process of automatically detailing the properties of the controls during the operation of a structure containing the data. Using data binding and writing, the code will create the connection quickly and accurately.

Conclusion

Low or high prices and trading momentum for profit taking covers a great deal of risk, where the decision to invest in stocks is based on recent purchases by other market participants. A drop in prices may also lead to new buyers coming to the market and pressure to buy may continue to push the price higher.

It is time for Morocco to adopt laws that regulate the trading industry, in order to close the door against the escort who presents himself as a broker and takes advantage of the situation to steal money from people.

The authorities should make greater efforts to understand the implications of the financial sector, detect fraud until risk management is improved and the applicability of laws is fact-based and jurisdictional, and achieve a more efficient and less costly financial system through more efficient treatment of credit risk information.

References

[1] Andrei A. Kirilenko and Andrew W. Lo (2013), Moore's Law versus Murphy's Law: Algorithmic Trading and Its Discontents, *Journal of Economic Perspectives*, 27(2):

51–72; Susan Athey (2017), "Beyond Prediction: Using Big Data for Policy Problems," *Science,* 355(6324): 483–485.

[2] In addition to the applications of AI and machine learning, there are a number of potential applications of distributed ledger technology (DLT), cloud computing and digital identity to regulatory reporting. These Reg. Tech applications are beyond the scope of this paper. See IIF (2017) for more detail.

[3] Christopher Matthews (2013), How Does One Fake Tweet Cause a Stock Market Crash?. http://business.time.com/2013/04/24/how-does-one-fake-tweet-cause-a-stock-market-crash/ Accessed on June 2014.

[4] P. Dwyer (1998), The 21st Century Stock Market, *Business Week*, Aug. 10, 1998.

[5] B. Arden (Ed.). (1980), *What Can be Automated?—The Computer Science and Engineering Research Study*. Cambridge, MA: The MIT Press.

[6] https://physics.stackexchange.com/questions/194309/where-does-particle-borrow-energy-from-to-tunnel.

[7] www.fsb.org/wp-content/uploads/P011117.pdf.

[8] Ethem Alpaydin (2014), *Introduction to Machine Learning*. MIT Press.

[9] Sean Jackson (2016), Big Data in Big Numbers – It's Time to Forget the 'three Vs' and Look at Real-world Figures. *Computing*, 18 February. www.computing.co.uk/ctg/opinion/2447523/big-data-in-big-numbers-its-time-toforget-the-three-vs-and-look-at-real-world-figures. Accessed 7 December 2016.

[10] https://ico.org.uk/media/for-organisations-data-protection.pdf.

[11] Simon Sharwood. Forget Big Data Hype Says Gartner as It Cans Its Hype Cycle. *The Register*, 21 August 2015.

[12] https://stockcharts.com/school/doku.php?id=chart_school:trading_harmonic_patterns.

[13] www.investopedia.com/articles/forex/11/harmonic-patterns-in-the-currency-markets.asp.

[14] N. T. Zung and N. T. Thien (2015), Reduction and Integrality of Stochastic Dynamical Systems. *Fundam. Prikl. Mat.* (in Russian) 20 (2015), no. 3, 213–249; English version: arXiv:1410.5492.

[15] http://trader.wikia.com/wiki/Momentum_(finance)

[16] N. Jegadeesh and S. Titman (1999), Profitability of Momentum Strategies: An Evaluation of Alternative Explanations. NBER Working paper #7159.

[17] K. Daniel, D. Hirschleifer and A. Subrahmanyam (1998), A Theory of Overconfidence, Self-Attribution, and Security Market under and Over-reactions. *Journal of Finance*, 53.

[18] J. Crombez (2001), Momentum, Rational Agents and Efficient Markets. *The Journal of Psychology and Financial Markets*, 2.

[19] http://ijesm.co.in/uploads/68/4975_pdf.pdf.

[20] https://dl.acm.org/citation.cfm?id=3234701.

[21] www.droitdunet.fr/les-differents-type-de-broker-dealing-desk-et-no-dealing-desk/.

[22] www.lesechos.fr/finance-marches/vernimmen/definition_spread.html.

[23] Deep learning is covered in a recent book by Bengio et al – Bengio (2016), Deep learning algorithms are also reviewed at length in Schmidhuber (2014), LeCun (2015); older reviews include Bengio (2012), Bengio (2009) and Graves (2012). Reinforcement Learning is covered in the classic text by Sutton and Barto (); newer treatments are available from David Silver, John Schulman and documentation of Open AI's gym framework.

[24] For data analysis in Python, see pandas (data manipulation), numpy/scipy (numerical computation and optimization), statsmodels/scikitlearn (time-series and machine learning) and pymc3 (Bayesian statistic).

15 Pattern to build a robust trend indicator for automated trading

Khalid Abouloula, Ali Ou-Yassine and Salah-ddine Krit

Introduction

To guide stock market trading, generally two widely used approaches to harvesting the information are used. Fundamental analysis is the most common approach to market research, addressing many external factors that are supposed to affect supply and demand in the market. The sources that can be obtained from the information include basic analysis: corporate annual reports, political and economic news, local and external events, and government policies. By carefully examining market supply and demand factors, it will be possible to predict changes in market conditions. So these changes are related to the volatility of the market price. Therefore, each trader must verify the accuracy of information about the currency market. Market trends are changing based on the news flow that is so complex today that it cannot be analyzed accurately. So we say that these changes are related to the volatility of the market price [1]. There are two factors that are composite in nature, and nobody can tell precisely the relative importance of their inter-relations that collectively lead to the final outcome [2].

One of the reasons for the fall of the stock market a few years ago that led to failure is the news of the decline in shares of dot-com companies, promoted by the fuss of the new digital economy. This indicates that angry discussions are still ongoing about the effectiveness of fundamental analysis, as well as other analysis.

The second popular approach is called the market forecasting technical analysis that operates in opposite principle from the other fundamental analysis. The underlying philosophy of this approach is that the market price also reflects we all know the factors at all times. So the price is already a strong performance indicator due to supply and demand for this particular market. Technical analysis is based solely on market prices themselves, rather than on another key factor outside the market. Traders who are only equipped with technical analysis assume that accurate analysis of daily price movement and a long term trend are all that is required to predict the direction of price for their trade [3].

Technical analysis

Forecasting market trends is a fresh field in the scientific research community using computing methods. For many years, traders have been hoping for some reliable decision-making tools that will help them to predict the market by relying on many commercial products and academic research models that have not been enough to achieve their goals. Some popular choices are genetic algorithms [4], support vector machines [5], and artificial neural networks [6]. They are used to analyze past financial data as far back as 20 years ago to try divining the market direction. The technical analysis is the study of the evolution of a market, mainly on the basis of graphs, in order to predict future trends. It is based on price evolution is the result of everything the rest, the variation of the course curves follow major and minor trends and the appearance of a frank trend encourages operators to act in the same direction as the trend.

Technical analysis reveals the regularly repeating market situations by two commonly used methods. Chart analysis is based on the graphical study of price and volume graphs. It is the appearance of certain graphic configurations that generates buying or selling signals. Charts are an important tool for many online traders and can play a key role in understanding technical analysis. You may get an overview of charts and figures and you are a bit surprised, but when you invest in stocks, commodities, indices or foreign exchange in the form of contract for differences CFDs, these charts can be your best friends. They provide you with information, give an insight into the long term topic, and can help you make trading decisions [7].

The story of market

The design of the strategies based on technical analysis gamble on a repetition of particular situations making it possible to predict the evolution of the courses in the periods to come. A mystic vocabulary is often used as resistance, support configurations and round figures meant to act as upward or downward psychological barriers.

We can anticipate that some exceptional peak (or other particular pattern) of the market trend that happen today, will one day happen again, just like how it did happen in history. For instance, in the 1997 Asian Financial Crisis [9], the Hang Sang Index in Hong Kong plunged from top to bottom (in stages 3 to 4); then about 10 years later, the scenario repeated itself in the 2008 Financial Crisis [10] with a similar pattern.

Mathematical study of trend indicators

Mathematical calculation is based on more information that aims to forecast the financials market direction [11] with named technical indicators which are organized into oscillators, volumes and trends indicators; we will focus on studying the trends indicators, explanations and equations.

Average directional index (ADI)

The ADI is an oscillator used to indicate strength or weak trend. It is an oscillator used to indicate strength or double direction and provides evidence of the current direction of the move and indicates only if there is a gain or strength. It is useful to judge if the market is moving sideways with a trend, however ADI comes in a combination of three elements: the ADI line itself and two other lines known as the positive direction index (+DI) and the negative direction index (−DI). These two additional components complement the ADI during cursor direction determination [12].

The ADI is a pointer between 0 and 100 points. A reading lower than 20 indicates a double trend. A bearish trend reading above 50 indicates a strong trend.

When the +DI crosses above the −DI, the price is higher in an uptrend and the ADI line measures the strength level above the 20 level below 50 when −DI is above +DI. Downwards ADI measures the strength of the falling trend.

Simple moving average

The simple moving arithmetic mean is the most popular arithmetic mean. It is the easiest method and is calculated as follows: if we want the five-day arithmetic mean, we will collect the five-day intervals and divide them by the number of days which is five [13], generally:

$$SMA = \Sigma \left(Close \; (i) \right) / N$$

where:
    Close (i) is the close price
    N is number of calculate periods.

Bollinger bands

The Bollinger bands index is a measure of the strength of the acceleration of price movements or the measurement of the degree of market instability in a period of time. The objective of this indicator is to find that there are moving and non-stationary levels of support and resistance that help to know the rebound areas in the market movement. Price volatility in the market is whether the market is quiet or there are sharp fluctuations in its movement [14].

When the price reaches the upper bar, it means that the assets are trading at a relatively high price and are at the peak of a buy. Consider selling the asset and expecting its price to fall again towards the central moving average bar.

When the price reaches the bottom bar, it means that the assets are trading at a relatively low price and are at the peak of the sell. Consider buying the asset and expecting its price to rise again towards the central moving average bar.

The Bollinger bands can also help you measure market volatility according to the distance between the upper and lower bands. The large distance between the bars indicates high volatility while the narrow distance indicates a low volatility.

Envelopes

Envelope index is one of the best indicators that determine the entry and exit points very accurately; also the indicator is characterized by high accuracy in monitoring the cases of breaking the trend and the reversal of the price situation during trading [15].

The index consists of two moving averages. One is moving higher than the price line where it represents the resistance point and should not be exceeded. The other moving average is moving below the price line as the support point that should not be crossed. It is worth mentioning that both lines move in parallel with the currency moving average. The sell signal is formed if the price approaches the resistance line, while the buy signal is formed if the price approaches the support line. To calculate the resistance and support steps required to determine the index, the following equations are used:

$$\text{Upper band} = \text{SMA} \left(\text{Close} \left(i \right) \right) \times \left[1 + K/1000 \right]$$
$$\text{Lower band} = \text{SMA} \left(\text{Close} \left(i \right) \right) \times \left[1 - K/1000 \right]$$

where:

Close (i) is close price

N is the period of averaging

K/1000 is the value of shifting from the average.

Standard deviation

Standard deviation is a measure that measures the volume of the current price movement of an asset, to predict future volatility in the price. It can help you determine whether volatility is likely to increase or decrease. The standard deviation index compares the movement of the current price to the historical price movement.

When the number of participants in the market rises, this indicates a rise in activity and price movement in the market, so the indicator can be easily interpreted. If the index of the standard deviation is low, prices are not active, it is logical to expect severe price volatility in the near future. If the value of the standard deviation index is high, price activity is likely to be as high as in the case of price breaks for support and resistance levels.

Process of trend indicators

The process is a random process which is a random mathematical function in most applications and probability models. The function is defined on a time domain or time period. In other cases, the function is defined on a space of space or a field of multidimensional space familiar and common in stock markets and exchange rate exchanges.

In mathematics, the fixed name may have two different meanings. It may indicate a fixed and well-defined mathematical object, and may also indicate a function. This constant is usually represented by a variable that does not depend on the main variable to study the problem.

General process structure of random walk is:

$$Xt+1 = Xt + \mathcal{E}t+1 \tag{15.1}$$

where:

X t is a time series
$\mathcal{E}t$ is white noise.

White noise is a process of null expectation E ($\mathcal{E}t$) and constant variance *Var* ($\mathcal{E}i$) = σ2. It is an independent identically distributed that refer to a statistical model for signals and signal sources, rather than to any specific signal [16]. So,

$$Xt+1 = Xt-1+\mathcal{E}t+\mathcal{E}t+1 \tag{15.2}$$

$$= Xt-2+\mathcal{E}t-1+\mathcal{E}t+\mathcal{E}t+1 \tag{15.3}$$

$$= Xt-3+\mathcal{E}t-2+\mathcal{E}t-1+\mathcal{E}t+\mathcal{E}t+1 \tag{15.4}$$

. . . .

$$= Xt-k+\mathcal{E}t+1+\mathcal{E}t+\mathcal{E}t-1+\mathcal{E}t-2+....+\mathcal{E}tk+1 \tag{15.5}$$

$$= Xt-k+\sum_{i=0}^{k-1}\mathcal{E}t-i \tag{15.6}$$

When: k = t

$$Xt+1 = X0+\sum_{i=0}^{t-1}\mathcal{E}t-i \tag{15.7}$$

Finally we obtain:

$$Xt = X0+\sum_{i=1}^{t}\mathcal{E}i \tag{15.8}$$

By demonstrated the random walk, we can use it to calculate the expectation and the variance from equation 15.1.

Expectation:

$$E(Xt) = E(X0+\sum_{i=1}^{t}\mathcal{E}i) \tag{15.9}$$

$$= E(X0)+E\left(\sum_{i=1}^{t}\mathcal{E}i\right) \tag{15.10}$$

We have X0 is a constant so E (X0) = X0.
So the results give:

$$E(Xt) = X0+E\left(\sum_{i=1}^{t}\mathcal{E}i\right) \tag{15.11}$$

And we can also write as:

$$= X0 + \left(\sum\nolimits_{i=1}^{t} E\left(\mathcal{E}i\right) \right) \tag{15.12}$$

Finally:

$$E\left(Xt\right) = X0 + E(\mathcal{E}1) + E(\mathcal{E}2) + E(\mathcal{E}3) + \ldots + E(\mathcal{E}t) \tag{15.13}$$

Or the principal condition of the random process is E $(\mathcal{E}t) = 0$.
We obtain the result as:

$$E\left(Xt\right) = X0 \tag{15.14}$$

So the expectation independent of time, it works with a stationary process.
Variance:

$$Var\ (Xt) = Var(X0 + \sum\nolimits_{(i=1)}^{(t)} \mathcal{E}i) \tag{15.15}$$

$$= Var(X0) + Var\ (\sum\nolimits_{(i=1)}^{(t)} \mathcal{E}i) \tag{15.16}$$

We have X0 is a constant so Var (X 0) = 0.
So the results give:

$$Var\left(Xt\right) = Var\left(\sum\nolimits_{i=1}^{t} \mathcal{E}i \right) \tag{15.17}$$

And we can also write as:

$$= \sum\nolimits_{i=1}^{t} Var(\mathcal{E}i) \tag{15.18}$$

The condition of the random process is Var $(\mathcal{E}i) = \sigma 2$.
Final:

$$Var\left(Xt\right) = t * \sigma 2 \tag{15.19}$$

The variance depends to the time, so does not work with a stationary process, we can do it by taking:

$$Xt = Xt - 1 + \mathcal{E}t \tag{15.20}$$

Switch to difference by adding X t − 1 left and right we have:

$$Xt - Xt - 1 = Xt - 1 - Xt - 1 + \mathcal{E}t \tag{15.21}$$

We obtain the order of integration:

$$\Delta X = \mathcal{E}t \qquad (15.22)$$

ΔX is the variation of X, so the chronological series are independent of time so it is now a stationary process.

Equation 15.8 is said to present a moving average process or a moving average representation for the simple building block provided by the noise. It is playing the role of building blocks for process with more complicated dynamics.

For this reason we were focused on what follows on process that are stationary or become so after a transformation.

Technological practices

Hybrid indicators that are a combine of the advantages the trend ones the precision of the oscillators, in this family, we analyzed the super-trend indicator as published on http://mql4.com by (Robinson, 2008) [18], processing described by (Kolier, 2010) [19], and evaluation of parameters for major FOREX by (Schmidt, 2011) [20].

Traders have recently proposed and developed the new super-trend indicators based on the two technical indicators for MetaTrader 4: ATR (Average true range) with a period of 5 and CCI (Commodity Channel Index) with a period of 50 [21].

On an uptrend CC (50) > 0: the values of the Super-Trend indicator are defined as the sum of: (ATR value) + (The high of last closed period).

In a falling market CC (50) < 0, the value gives the difference: (the low of the last closed period) − (ATR value).

With using the history centre of the MetaTrader 4 for six major pairs and for the time period as shown in Table 15.1.

All the next works will be done with MetaTrader 5 platform, which is an interface of development relative to Meta Quote language MQL5, source code developed in. mql5 files, and the result requires compiling before to be used.

Table 15.1 History centre of the MetaTrader 4 for six major pairs

Currency pair	Oldest time point
EUR/USD	01/1999
GBP/USD	01/1999
EUR/GBP	01/1999
GBP/CHF	01/1999
AUD/USD	06/2003
EUR/AUD	12/2006

ALGORITHM: super-trend indicators

```
void OnInit()
  {
  if(InpCCIPeriod<=0)
     {
     ExtCCIPeriod=50;
     printf("Incorrect    value    for    input    variable
     InpCCIPeriod=%d. Indicator will use value=%d for
     calculations.",InpCCIPeriod,ExtCCIPeriod);
     }
  else ExtCCIPeriod=InpCCIPeriod;
  SetIndexBuffer(0,ExtCCIBuffer); SetIndexBuffer(1,
  ExtDBuffer,INDICATOR_CALCULATIONS); SetIndexBuff
  er(2,ExtMBuffer,INDICATOR_CALCULATIONS); SetInd
  exBuffer(3,ExtSPBuffer,INDICATOR_CALCULATIONS);
  IndicatorSetString(INDICATOR_SHORTNAME,"CCI("+string
  (ExtCCIPeriod)+")");
  PlotIndexSetInteger(0,PLOT_DRAW_BEGIN,ExtCCI
  Period-1);
  IndicatorSetInteger(INDICATOR_DIGITS,2);
  }

void OnInit()
  {
  if(InpAtrPeriod<=0)
     {
     ExtPeriodATR=5;
     printf("Incorrect    input    parameter    InpAtrPer-
     iod = %d. Indicator will use value %d for calcula
     tions.",InpAtrPeriod,ExtPeriodATR);
     }
  else  ExtPeriodATR=InpAtrPeriod; SetIndexBuffer(0,Ext
  ATRBuffer,INDICATOR_DATA); SetIndexBuffer(1,ExtTRBuf
  fer,INDICATOR_CALCULATIONS);
  IndicatorSetInteger(INDICATOR_DIGITS,_Digits);
  PlotIndexSetInteger(0,PLOT_DRAW_BEGIN,InpAtrPeriod);
  string short_name="ATR("+string(ExtPeriodATR)+")";
  IndicatorSetString(INDICATOR_SHORTNAME,short_name);
  PlotIndexSetString(0,PLOT_LABEL,short_name);
  }
     {
     IndicatorRelease (m_handle);
     m_handle = INVALID_HANDLE;
     return (false);
     }
  return (true);
  }
```

Once switched the algorithm in MetaTrader 5, the approach is different by the fact that it produces calculations on several time periods.

With the addition of that observation, the super-trend indicator does neither price forecasting nor predicting any market movement. Some studies show the existence of ecologies of high-frequency trading practices [22], others use discourse analysis to show that automated trading is more like an arrangement of different epistemic communities that differ according to the sector being apprehended and whose practices depend largely on the stakeholder concerned [23]. The concern of a construction of meaning made by the actors of the financial markets then appears as the common denominator of these approaches. Market finance contributions have mainly resulted in standard approaches based on asset price valuation and investor behaviour [24]. The most empirical approaches that consider the speed associated with new trading technologies are very contrasting in their results. The trend will be calculated on the average price difference for a fast and a slow period and expressed in percentage from the close price of the current bar. To be able to correctly read price action, trends and trend direction, we will introduce the most effective ways to analyze a chart.

We will use and mixed character between the five trend indicators ADI, SMA, BB, envelopes and SD linked by moving average trend which is the base in a common algorithm to implement and integrate the performance progression of script and development of trading, is equipped with some mechanisms that limit the losses. Before giving more details about our algorithm, we give in the following some preliminary necessary on "the trend methods" and "Concept of trend trading".

The trends methods

The idea of using trend indicator is the most important concept in technical analysis; trend is an integrated environment that meets the needs of traders through the speed and performance of the program and advanced tools for follow-up and unique control make the goals accessible. The trend can be described as the embodiment of suggestions and developments over the years of hard work and continuous development.

Trend definition

A trend is a certain trend of price action, and trend or trend behaviour is evaluated when technical analysis is performed. Trends may be bearish (also known as bear trends), and may be bullish (known as bull trends) or a horizontal direction (empty directions) [25], as a rule when the trend is bearish, it is recommended to open the short positions, and when the trend is upward, it is advisable to open the buying positions, when there is no specific direction, it is best not to do any operations [26].

The index is expressed in a number of points, where the average value reflects the securities listed in the market either in whole or in part, if the index is calculated in a specific sample of the market, the result is comprehensive for all securities in the market, where the volume of liquidity of the paper in the market and its activity. To choose the securities in the calculation of the index, and so we can know the performance of a sector by calculating the index of this sector by the same mechanism of calculating the market index.

The calculation of the index depends on a set of assumptions that fit the nature of both the market and the securities listed in it so that we can calculate more than one market index by changing one of the assumptions as the difference of indicators for the market is due to different assumptions and not on the actual trading results such as prices and quantities traded and restricted.

Types of trends

Those trends are uptrends, downtrends and sideways, however this does not mean that it moves in straight lines. With direction, the movement ranges between ascending and descending. It is like a zigzag. It consists of highs and lows during the price movement, called peaks and troughs.

Candles

In order to analyze the Forex market, it is important to understand the Japanese candle building. The main building of the Japanese candlestick chart is the candle. Understanding how to read candles is essential to determine if a financial bond follows a specific direction. Candles may look a bit confusing, but as soon as we understand them, we'll see that they represent a simple and effective way to track the value of bonds available, and that's what we'll be working on. A bar is a visual representation of a block of time or price movement. Bars usually contain information such as open, close, high and low, volume and time. Different types of bars exist and are used by chartists, such as candlesticks (Figure 15.1).

The candle may be upward or downward; the green candle indicates that the price rises while the red candle indicates the opposite. It is important to distinguish between the main body of the candle and the lines that extend from it.

Uptrends

The price is said to be bullish when the consecutive lows higher than the previous bottoms, as shown below. The bottom 3 is higher than the bottom 1 and this puts the bullish trend, which is confirmed only at point 5, the third high bottom [27], as shown in Figure 15.2.

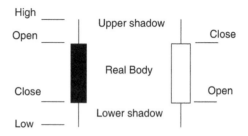

Figure 15.1 Candlesticks

Figure 15.2 An example of an uptrend

Figure 15.3 An example of a downtrend

Downtrend

The price is said to be in a downward trend at successive highs below previous highs, as shown below, you can see that the fourth point is below the 2 point, and we note again 4 is less than 2 [28], as shown in Figure 15.3.

Trend lines

Trend lines are lines drawn at a higher or lower angle. Trend lines are used to give signals about the current trend and also give other signals when the trend changes. These lines can be used as support and resistance and thus offer opportunities to open and close trading positions [29].

Channels

Channels are another tool used in technical analysis that can be used to identify good places to buy or sell. They represent both the tops and bottoms of channels of potential resistance or support areas. To form an ascending channel, simply draw a parallel line in the same corner of the ascending trend line and move this line to where it touches the last peak and this should be done at the same time as the trend line. To form a (bearish) channel, simply draw a parallel line in the same corner of the descending trend line and move this line to where it touches the bottom and this should be done at the same time as the trend line. When prices arrive for the lower trend line, they can be used as a buy zone [30]. When prices reach the top trend line, they can be used as a selling area as shown at Figure 15.4.

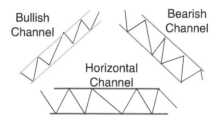

Figure 15.4 Three types of price channels

Concept of trend trading

Algorithmic trading

Algorithmic trading consists of using a software program to make trades on a market using some special algorithm. After analysing price, the algorithm makes decisions about the trade: direction, timing, quantity, order type and so forth. Algorithmic trading is used by banks and hedge funds for example. Due to free market platform such as Meta Trader and brokers with low capital requirements, almost anyone can participate in algorithmic trading nowadays [31]. Below is a list of some pros and cons of algorithmic trading.

Advantages of algorithmic trading

Automate of the asset selection process. This allows the entire process to be reduced to step-by-step execution of instructions of the trading strategy by a certain program. That means that your entire trading strategy can be represented in the form of a trading robot, which itself will make operations and search for entry points. Absence of a psychological factor, the robot is not subject to emotions, and therefore will trade only based on an algorithm without deviations to the left or to the right. Diversification of the risks of manual trading, if you will trade on one account yourself, and entrust the second account to the robot, this will reduce the risks by diversifying the fix API trading.

Disadvantages of algorithmic trading

Prohibition of the broker companies to conduct trading with the help of trading robots. This trading method is also limited by providing negative values. However, this can be circumvented with the help of auxiliary programs that mask the robot trading. Not all trading strategies can be automated. There are many formalized strategies that are very difficult to automate in view of the author's features or subjective analysis of the fix API trader.

Trading program

Trading program an idea to refers trading system that automatically submits to a given exchange. Depending on trading frequency the speed of data may have a significant influence whether the trading system is profitable or not. Automated system trading consists of many all of which must be taken into consideration when building a reliable system. Trading program must have a reliable market data feed.

Alpha model

This data is analyzed by algorithms to find if there are currently such market conditions, which would have been likely profitable trading opportunities according to back test [32]. Alpha is often the active return on investment, and measures the performance of the investment versus the market index used as a benchmark, as it often represents the movement of the market as a whole, and the Alpha Fund is the excess fund revenue that relates to the return of the benchmark.

Risk model

Risk management should be taken into account to identify risks in a portfolio. We work on risk modelling to use econometric techniques within a broader range. Therefore, any prices displayed by a chart provider are indicative prices rather than actual trading prices [33].

Transaction cost model

That the cost of access to the centre is first result of the liquidity factor and prices offered through the providers of liquidity in the market and second from the commission, where the transactions are based on the nature of the costs borne by companies during the transaction process [34].

We resume the trading architecture in Figure 15.5.

The theory of efficient markets introduces some basic technical analysis tools for estimating trading signals related to trend, consequently, the speed itself is very crucial and it has suggested that any delay from 10 ms and 1 s leads to a statistically significant decrease in performance.

Software

Numerous software platforms for algorithmic trading exist. A couple of in this study will be introduced. In addition to platforms designed solely for trading, Software packages can be used for algorithmic development as well as actual execution of orders.

Meta Trader, Supports Forex, CFD and futures markets are the software is a package of both server and client components. The Meta Trader Client Terminal is used for technical analysis, trading and algorithmic trading for development.

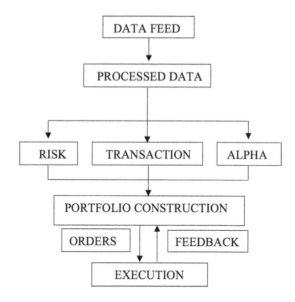

Figure 15.5 The trading architecture

Result and discussion

It was necessary to understand, analyze and identify trends so that we can consolidate them in one direction according to the suggested steps in this paper, in order to be able to choose the best trends that will help us to read sound market volatility. Add more than one pointer to the chart for an in-depth view of a particular tool, but always remember to adapt to time frames. In this section we show the results obtained using the simple moving average.

Code of SMA Trend:

```
double SMA (const int position, const int period,
const double & price [])
  {
  double result=0.0;
  if (position>=period-1 && period>0)
    {for (int i=0 ; I < period ; i++)
      result+=price[position-i];
        result/=period;
    }
  Return (result);
  }
```

Using the SMA Trend with zero period, the position coincides with the graphic trading, so with using the high and low simple moving average we envelope the trading positions.

We use also the envelope and the Bollinger Bands trends in the same trade with we see what we want as shown in flowing,

The results show that our model perform better than double trend indicator algorithm, in our experiment simulates existing market securities, we see that the resulting equilibrium price of the optimal security is a strip et the equilibrium price of the optimal security at different points in time, it is not true to rely on only one indicator, but rely on more than one index in order to get a true reading free of false or false signals. In short, we can use the five trend indicators based on a moving average trend in one algorithm to be implemented and include progress-ing the performances of script trading with development as following program:

ALGORITHM: Trends

```
bool iADI :: Create (const string symbol, const ENUM_
TIMEFRAMES period,const int ma_period)
  {
  if (!SetSymbolPeriod(symbol,period))
    return (false);
  m_handle = iADI (symbol, period, ma_period);
  if (m_handle==INVALID_HANDLE)
    return (false);
  if (!Initialize (symbol, period, ma_period))
    {
    IndicatorRelease (m_handle);
    m_handle = INVALID_HANDLE;
    return(false);
    }
  return(true);
  }

bool iBands :: Create (const string symbol, const ENUM_
TIMEFRAMES period, const int ma_period, const int ma_
shift, const double deviation, const int applied)

  {
  if (!SetSymbolPeriod(symbol, period))
    return(false);
  m_handle=iBands(symbol, period, ma_period, ma_shift,
  deviation, applied);
  if (m_handle==INVALID_HANDLE)
    return (false);
```

```
if(!Initialize(symbol,period,ma_period,ma_shift,
deviation,applied))
   {
    IndicatorRelease (m_handle);
    m_handle = INVALID_HANDLE;
    return (false);
   }
 return (true);
 }


bool iEnvelopes :: Create (const string symbol, const
ENUM_TIMEFRAMES period, const int ma_period, const int
ma_shift,const  ENUM_MA_METHOD  ma_method,  const  int
applied,const double deviation)
  {
  If (!SetSymbolPeriod (symbol, period))
    return (false);
  m_handle=iEnvelopes(symbol, period, ma_period, ma_
  shift, ma_method, applied, deviation);
  if (m_handle == INVALID_HANDLE)
    return (false);
  if (!Initialize(symbol, period, ma_period, ma_shift,
  ma_method, applied, deviation))
    {
     IndicatorRelease (m_handle);
     m_handle = INVALID_HANDLE;
     return (false);
    }
  return (true);
  }


bool iMA :: Create(const string symbol, const ENUM_TIME-
FRAMES period, const int ma_period,const int ma_shift,
const ENUM_MA_METHOD ma_method, const int applied)
  {
  if (!SetSymbolPeriod (symbol, period))
    return (false);
  m_handle = iMA (symbol, period, ma_period, ma_shift,
  ma_method, applied);
  if (m_handle == INVALID_HANDLE)
    return (false);
  if (!Initialize (symbol, period, ma_period, ma_shift,
  ma_method, applied))
```

```
  {
    IndicatorRelease (m_handle);
    m_handle = INVALID_HANDLE;
    return (false);
  }
  return (true);
}

bool iStdDev :: Create (const string symbol, const
ENUM_TIMEFRAMES period, const int ma_period,const int
ma_shift, const ENUM_MA_METHOD ma_method, const int
applied)
  {
  if (!SetSymbolPeriod (symbol, period))
    return (false);
    m_handle = iStdDev (symbol, period, ma_period, ma_
    shift, ma_method, applied);
  if (m_handle == INVALID_HANDLE)
    return (false);
  if (!Initialize (symbol, period, ma_period, ma_shift,
  ma_method, applied))
    {
    IndicatorRelease (m_handle);
    m_handle = INVALID_HANDLE;
    return (false);
    }
  return (true);
  }
```

We proposed this model, through which we applied an automatic based on the accuracy of the prediction of the time series, by historical monitoring of trends employed for profitable trading.

Conclusion

In order to determine the date of purchase or sale, there are some specific rules for this which is known as market trading technology. In this work, we improved the algorithm that was based only on the diodes in the market directive by adding three trends to the previous two. Among them one of the transactions as studied, which gave important results and is able to generate profit from trading in the stock market. This algorithm also showed support for novice traders because they do not share market volatility in large ranges with good profit through the pro-posed trading strategy.

References

[1]   C. Robertson, S. Geva, and R. Wolff (2006), "What Types of Events Provide the Strongest Evidence that the Stock Market Is Affected by Company Specific News," *Proceedings of the Fifth Australasian Conference on Data Mining and Analytics,* Vol. 61, Sydney, Australia.

[2]   www.cis.umac.mo/~fstasp/paper/jetwi2011.pdf.

[3]   http://ijesm.co.in/uploads/68/4975_pdf.pdf.

[4]   D. Fuente, A. Garrido, J. Laviada, and A. Gomez (2006), "Genetic Algorithms to Optimise the Time to Make Stock Market Investment," *Genetic and Evolutionary Computation Conference,* Seattle, USA, pp. 1857–1858.

[5]   Y. Sai, Z. Yuan, and K. Gao (2007), "Mining Stock Market Tendency by RS-Based Support Vector Machines," *The 2007 IEEE International Conference on Granular Computing, Silicon Valley,* USA, pp. 659–664.

[6]   S. Xu, B. Li, and Y. Shao (2008), "Neural Network Approach Based on Agent to Predict Stock Performance," *IEEE International Conference on Computer Science and Software Engineering,* Vol.1, Wuhan, China, pp. 1223–1225.

[7]   www.ocf.berkeley.edu/~jml/decal/techanalysis.pdf.

[8]   Wikipedia, 1997 Asian Financial Crisis, Available at http://en.wikipedia.org/wiki/1997_Asian_Financial_Crisis, last accessed on July 3, 2012.

[9]   Wikipedia, Financial Crisis of 2007–2010, Available at http://en.wikipedia.org/wiki/Financial_crisis_of_2007, last accessed on July 3, 2012.

[10]   John J. Murphy (1999), *Technical Analysis of the Financial Markets: A Comprehensive Guide to Trading Methods and Applications (2nd ed).* New York [u.a.]: New York Inst. of Finance.

[11]   J. Welles Wilder, Jr. (June 1978), *New Concepts in Technical Trading Systems.* Greensboro, NC: Trend Research. ISBN 978–0894590276.

[12]   www.metatrader5.com/en/terminal/help/indicators/trend_indicators/ma.

[13]   (www.Bollingerbands.com) second paragraph, centre column.

[14]   The E-Book of technical Market Indicators 2.0 / Complex Technical Analysis Mode Simple/How to build a rational decision making framework (systematic trading model) based on different kinds of technical market indicators.

[15]   https://en.m.wikipedia.org/wiki/White_noise.

[16]   J. Robinson (2008), Code Base MQL4, Super Trend Source Code. Retrieved from http://codebase.mql4.com/.

[17]   L. Kolier (2010), How Super Trend Works. Retrieved 08, 2011, from http://kolier.li/indicator/how-supertrend-mq4-works-logic-of-supertrend-indicator-created-by-jason-rebinson-jnrtradin.

[18]   www.vtad.de/sites/files/forschung/Schmidt_2011_Evaluation_of_params_super trend_daily_VTAD.pdf.

[19]   http://dewinforex.com/forex-indicators/supertrend-hybrid-indicator-main-direction-of-the-market.html.

[20]   D. Mac Kenzie (2014), A Sociology of Algorithms: High-Frequency Trading and the Shaping of Markets, Manuscript non-public.

[21]   R. Seyert (2016), "Bugs, Predations or Manipulations? Incompatible Epistemic Regimes of High-Frequency Trading," *Economy & Society,* Vol. 45, No. 2, pp. 251–277.

[22] B. Beaufils, O. Brandony, L. Ma, and P. Matthieu (2009), " Simuler pour comprendre : un éclairage sur les dynamiques de marchés financiers à l'aide des systèmes multi-agents," *Systèmes d'Information et Management*, Vol. 14, No. 4, pp. 51–70.

[23] www.investopedia.com/university/technical/techanalysis3.asp.

[24] www.metatrader5.com/en/terminal/help/indicators/trend_indicators/envelopes.

[25] www.investopedia.com/university/technical/techanalysis11.asp.

[26] www.investopedia.com/terms/d/downtrend.asp#ixzz55kyud0fv.

[27] http://stockcharts.com/school/doku.php?id=chart_school:chart_analysis:trend_lines.

[28] www.bloomberg.com/research/stocks/private/snapshot.asp?privcapId=50174746.

[29] www.linkedin.com/pulse/pros-cons-algorithmic-trading-julia-fischer/.

[30] www.nomura.com/events/global-quantitative-equity-conference/resources/upload/ruy_alves.pdf.

[31] On the other hand, intra-day trading will typically increase the portfolio risk relative to close-to-close positions because trading positions are typically cut down toward the close of the day.

[32] www.businessdictionary.com/definition/transaction-cost-theory.html.

Index

Note: Page numbers in *italics* indicate a figure and page numbers in **bold** indicate a table on the corresponding page.